Advances in Discrete-Time Sliding Mode Control

Theory and Applications

Advances in Discrete-Time Sliding Mode Control

Theory and Applications

Ahmadreza Argha
Steven Su
Li Li
Hung Tan Nguyen
Branko George Celler

CRC Press is an imprint of the
Taylor & Francis Group, an **informa** business

CRC Press
Taylor & Francis Group
6000 Broken Sound Parkway NW, Suite 300
Boca Raton, FL 33487-2742

First issued in paperback 2020

ISBN-13: 978-0-367-57141-2 (pbk)
ISBN-13: 978-1-138-30027-9 (hbk)

Version Date: 20180515

Visit the Taylor & Francis Web site at
http://www.taylorandfrancis.com

and the CRC Press Web site at
http://www.crcpress.com

Ahmadreza Argha dedicates this book to his wife Naghmeh Akhtar.
Steven W. Su dedicates this book to his family.
Li Li dedicates this book to his family.
Hung T. Nguyen dedicates this book to his family.
Branko G. Celler dedicates this book to his family.

Contents

List of Figures xi

List of Tables xv

Symbols xvii

Preface xix

Contributors xxv

1 Introduction 1

1.1 Continuous-time SMC . . . 2
1.2 Regular form-based DSMC . . . 20
1.3 Summary . . . 22

I LMI-Based Discrete-Time Sliding Mode Control 25

2 LMI-based SF DSMC 27

2.1 Introduction . . . 27
2.2 Problem Formulation . . . 29
2.3 Design of Discrete-time SMC . . . 31
2.3.1 Variable structure discontinuous control considerations . . . 32
2.3.1.1 Using upper and lower bounds of disturbance in the controller: $\mathscr{C}_1$ . . . 33
2.3.1.2 Using $f_i^+ \pm \frac{f_i^-}{2}$: $\mathscr{C}_2$ . . . 34
2.3.2 Design of a robust sliding surface . . . 34
2.3.3 Characterizing system state boundedness . . . 37
2.4 Disturbance Estimate in Control Law . . . 39
2.5 Simulation Results . . . 41
2.6 Conclusions . . . 49

3 LMI-based output feedback DSMC 51

3.1 Introduction . . . 51
3.2 Problem Formulation . . . 53

3.3 Observer-Based Output Feedback DSMC 54
3.3.1 Stability analysis 56
3.3.2 Characterizing the system state boundedness 59
3.3.3 Discussions 61
3.4 Simulation Results 62
3.5 Conclusions 66

II DSMC for NCSs Involving Packet Losses 67

4 NCSs with measurement packet losses 69

4.1 Introduction 69
4.2 Problem Formulation and Preliminaries 71
4.3 Stochastic Sliding Mode Control 73
4.4 Variable Structure Controller Considerations 80
4.5 Simulation Results 81
4.6 Conclusions 86

5 NCSs with actuation and measurement packet losses 87

5.1 Introduction 87
5.2 Problem Formulation and Preliminaries 88
5.2.1 Problem statement 88
5.3 Stochastic Sliding Mode Control 91
5.3.1 Designing the sliding function subject to consecutive packet losses 92
5.3.2 Stability analysis 93
5.4 Numerical Examples 103
5.4.1 Example 1 103
5.5 Conclusions 104

III Sparse Sliding Mode Control for Large Scale NCSs 111

6 Sparse DSMC for NCSs 113

6.1 Introduction 114
6.2 Problem Formulation and Preliminaries 116
6.2.1 Preliminaries 117
6.2.2 State and disturbance observer 119
6.3 Spatially Decentralized Sliding Mode Control 120
6.4 Stability Analysis 122
6.5 Sparsifying the Control Network Structure 127
6.6 Numerical Examples 129
6.6.1 Example 1 129
6.6.2 Example 2 131
6.6.3 Example 3 138

6.7 Conclusions 138

7 Optimal sparse SMC for NCSs **141**

7.1 Introduction 142
7.2 Problem Formulation and Preliminaries 143
7.3 Optimal Structured SMC Design Problem 147
7.3.1 $\mathscr{H}_2$ based optimal structured static output feedback 147
7.3.2 Stability analysis of sliding mode dynamics 149
7.4 Sparsification of the Control Network 150
7.5 Numerical Examples 152
7.5.1 Example 1 152
7.5.2 Example 2 153
7.5.2.1 Comparison 1 153
7.5.2.2 Comparison 2 155
7.6 Solving LQ SOF Problem 155
7.7 Reweighted ℓ_1 Minimization Algorithm 157
7.8 Conclusions 157

IV DSMC for Two-Dimensional Systems **159**

8 DSMC for 2D systems **161**

8.1 Introduction 161
8.2 Problem Formulation 163
8.2.1 New 1D form of 2D first FM model 163
8.3 DSMC for 1D Discrete Vector Form 165
8.3.1 Direct method to find control law 168
8.4 Simulation Results 168
8.5 Conclusions 170

9 Controllability analysis of 2D systems **171**

9.1 Introduction 171
9.2 WAM Model of the First FM Model 172
9.2.1 Controllability analysis of the WAM model 174
9.3 Controllability Analysis of the New Model in 8.2.1 179
9.3.1 Notion of local controllability for 2D systems 179
9.3.2 Directional controllability with respect to $\{j\}$-direction . . . 180
9.3.3 Directional controllability with respect to $\{i\}$-direction . . . 182
9.3.4 Directional minimum energy control input 183
9.4 Numerical Example 183
9.5 Conclusions 185

V Integral DSMC for Heart Rate Regulation **187**

10 HR regulation during cycle-ergometer exercise **189**

10.1 Introduction 190
10.2 Methods 193
10.2.1 Equipment and data acquisition system 193
10.2.2 HR profile 194
10.2.3 Control system 195
10.2.3.1 Actuator-based event-driven PID controller 195
10.2.3.2 Actuator-based event-driven adaptive ISMC . . . 196
10.2.3.3 Auditory converter 199
10.2.3.4 The proposed control mechanism in summary . . 199
10.2.3.5 Two novel anti-windup mechanisms 200
10.2.3.6 Relay controller strategy 201
10.2.4 Tuning the PID controller gains 202
10.3 Results 202
10.4 Discussion 205
10.4.1 Necessity of the project 205
10.4.2 Discussion of the control system 206
10.4.3 Discussion of the results 212
10.5 Conclusions 213

Bibliography **215**

Index **227**

List of Figures

1.1 Evolution of system state using LQ regulator. 3
1.2 Evolution of system state obtained by applying SMC in (1.8) with $M = -1.8875$ and $\rho = 4$ to the system 1.1. 5
1.3 Evolution of switching function for the system 1.1 using SMC in (1.8) with $M = -1.8875$ and $\rho = 4$. 6
1.4 Evolution of switching function (zoom) for the system 1.1 using SMC in (1.8) with $M = -1.8875$ and $\rho = 4$. 7
1.5 Control effort with SMC in (1.8) with $M = -1.8875$ and $\rho = 4$. . . 8
1.6 Phase plot for the system (1.1) using SMC in (1.8) with $M = -1.8875$ and $\rho = 4$. 9
1.7 Evolution of system state obtained by applying SMC in (1.12) with $M = -1.8875$, $\rho = 4$ and $\varepsilon = 0.1$ to the system (1.1). 11
1.8 Evolution of system state (zoom) obtained by applying SMC in (1.12) with $M = -1.8875$, $\rho = 4$ and $\varepsilon = 0.1$ to the system (1.1). . 12
1.9 Evolution of switching function for the system (1.1) using SMC in (1.12) with $M = -1.8875$, $\rho = 4$ and $\varepsilon = 0.1$. 13
1.10 Evolution of switching function (zoom) for the system (1.1) using SMC in (1.12) with $M = -1.8875$, $\rho = 4$ and $\varepsilon = 0.1$. 14
1.11 Control effort with SMC in (1.12) with $M = -1.8875$, $\rho = 4$ and $\varepsilon = 0.1$. 15
1.12 Evolution of system state obtained by applying SMC in (1.13) and (1.12) with $M = -1.8875$, $\rho = 4$, $\varepsilon = 0.1$ and $\varphi = -1$ to the system (1.1). 16
1.13 Evolution of switching function for the system (1.1) using SMC in (1.13) and (1.12) with $M = -1.8875$, $\rho = 4$, $\varepsilon = 0.1$ and $\varphi = -1$. 17
1.14 Control effort with SMC in (1.13) and (1.12) with $M = -1.8875$, $\rho = 4$, $\varepsilon = 0.1$ and $\varphi = -1$. 18
1.15 Phase portrait for the system (1.1) using SMC in (1.13) and (1.12) with $M = -1.8875$, $\rho = 4$, $\varepsilon = 0.1$ and $\varphi = -1$. 19

2.1 Signal $f_i(k)$. 32
2.2 Plant: Single dimensional motion of a unit mass. 42
2.3 Results of linear controller. 43
2.4 Results of linear controller and using mean value of disturbance in DSMC. 44
2.5 Results of controller $\mathscr{C}_1$. 45

2.6 Results of controller $\mathscr{C}_2$. 46
2.7 Results of the controller in [101]. 47
2.8 Results of controller $\mathscr{C}_3$. 48

3.1 Results of ODSMC with disturbance estimator (3.10). 63
3.2 Results of applying ODSMC in (3.47). 65

4.1 NCS structure. 72
4.2 Bernoulli sequence $\alpha(k)$. 82
4.3 Results of applying DSMC in (4.43). 83
4.4 Results of applying DSMC in (4.45). 84
4.5 Results of applying linear controller. 85

5.1 NCS structure 89
5.2 Control effort u_1. 105
5.3 Control effort u_2. 106
5.4 Trajectories of the sliding function. 107
5.5 Trajectories of the system output and its estimate. 108
5.6 Trajectories of the system output and its estimate. 109
5.7 Bernoulli sequences (a) $\alpha(k)$, (b) $\beta(k)$. 110

6.1 Three coupled inverted pendulums system. 130
6.2 Trajectories of the system state with fully distributed structure. . . . 132
6.3 Deviation from the sliding surface with fully distributed structure . 133
6.4 Control efforts with fully distributed structure. 134
6.5 Exogenous disturbances and disturbance estimator outputs with fully distributed structure. 135
6.6 Trajectories of the system state with structure $\Gamma^{\star}$. 136
6.7 Deviation from the sliding surface with structure $\Gamma^{\star}$. 137
6.8 Control efforts with structure $\Gamma^{\star}$. 137

8.1 The system state x_1. 169
8.2 The system state x_2. 169
8.3 The control law u. 170

9.1 2D system state 184

10.1 Mechanism of the proposed control system. 191
10.2 Block diagram for the HR regulation system during cycling exercise using non-model-based control schemes. 192
10.3 Block diagram for the HR regulation system during cycling exercise using model-based control schemes. 192
10.4 Nonin 4100 Pulse Oximeter 194
10.5 Reed switch and adjustable time delay parameter 195
10.6 Dynamic profile mechanism used as anti-windup 201

10.7 HR profile tracking during cycling using event-driven PID controller. . . . 203
10.8 HR profile tracking during cycling using conventional PID controller and fixed-rate biofeedback 204
10.9 HR profile tracking during cycling using the relay controller 204
10.10 HR profile tracking during cycling using event-driven ISMC and damped RLS 205
10.11 Cycle-ergometer exercising system 207
10.12 Mechanism of a sensor-based event driven control system using state observer. T_{s_i} is the varying output sampling rate and T_{sys} is the sampling rate of the system. 208
10.13 Mechanism of a sensor-based event driven control system using spatial domain instead of time domain. T_{s_i} is the varying output sampling rate and control signal update rate. 209
10.14 Mechanism of an actuator-based event driven control system. T_s is the constant output sampling rate and T_{c_i} is the varying event occurrence rate (pedaling rate). 210

List of Tables

6.1 Comparison of E_{rms} for different structures 138

Symbols

Symbol Description

Symbol	Description
I_m	Identity matrix of size $m \times m$
A^T	Transpose of the matrix A
$\lambda(A)$	Eigenvalues of the matrix A
$\lambda_{\max}(A)$	Maximum eigenvalue of the matrix A
$\lambda_{\min}(A)$	Minimum eigenvalue of the matrix A
$\mathbb{R}$	Collection of real numbers
$\|x\|$	2-Norm of the vector x defined as $\sqrt{x^T x}$
$\|A\|$	2-Norm of the matrix A defined as $\max \frac{\|Ax\|^2}{\|x\|^2}$
$\lvert x(t) \rvert$	Absolute value of the scalar x at time t
$\|x(t)\|$	2-norm of the vector x at time t
$\dot{x}(t)$	Time derivative of the vector $x(t)$, i.e. $\dot{x}(t) = \frac{dx(t)}{dt}$
$[\Sigma_{ij}]_{r \times r}$	is a block matrix with block entries Σ_{ij}, $i = 1, \cdots, r$, $j = 1, \cdots, r$
$\operatorname{diag}[\Sigma_{ii}]_{i=1}^{r}$	is a block-diagonal matrix with block entries Σ_{ii}, $i = 1, \cdots, r$
$\operatorname{col}(v_i(k))_{i=1}^{r}$	denotes a block-vector with block entries $v_i(k)$, $i = 1, \cdots, r$
$\{\circ\}$	denotes an operator for $\Xi = [\xi_{ij}]_{h \times h}$ in which $\xi_{ij} \in \mathbb{R}$ and $W = [W_{ij}]_{h \times h}$ in which $W_{ij} \in \mathbb{R}^{r_i \times s_j}$ such that $\Xi \circ W = [\xi_{ij} W_{ij}]_{h \times h}$
$\otimes$	Kronecker product
SMC	Sliding Mode Control
VSDCS	Variable Structure Discontinuous Control Strategy
CSMC	Continuous-Time Sliding Mode Control
DSMC	Discrete-Time Sliding Mode Control
QSM	Quasi Sliding Mode
LMI	Linear Matrix Inequality
PIO	Proportional Integral Observer
NCS	Networked Control System
MIMO	Multiple Input, Multiple Output
SISO	Single Input, Single Output
OSMC	Output Based Sliding Mode Control
OCSMC	Output Based Continuous-Time Sliding Mode Control
ODSMC	Output Based Discrete-Time Sliding Mode Control
RMS	Root Mean Squares
RLS	Recursive Least Squares
ISMC	Integral Sliding Mode Control
2D	Two-Dimensional
1D	One-Dimensional
FM	Fornasini and Marchesini
RM	Roesser Model

WAM	Wave Advanced Model	HR	Heart Rate
PID	Proportional, Integral and Derivative		

Preface

Sliding mode control (SMC) commenced in the Soviet Union in the late 1950s, but this new control technique was not published until the publications [70] and [113]. Then, the sliding mode research community expanded quickly and the number of publications on this control framework grew correspondingly. Due to the fact that SMC relies on an infinite switching frequency of the input signal, it is inherently a continuous-time control strategy. However, the infinite switching is not achievable in real applications, especially for discrete-time controllers whose input signal can only be varied at the sampling instances. This fact limits the switching frequency to the discrete-time system's sampling frequency. It is worth noting that in a number of applications the assumption of an infinite switching frequency can be relatively justified. In the case that the sampling rate is much faster than the dynamics of the system under control, the influence of the bounded switching frequency will be confined. It is thus a usual approach to design sliding mode controllers in the continuous-time domain, even if the system is computer-aided-controlled [149], regarded as continuous-time sliding mode controller (CSMC), since it is designed according to a continuous-time model of the system, regardless of the sampling issue. However, the effectiveness of the obtained controller will, in addition to many other parameters, strongly depend on the sampling frequency. It means that the faster sampling is performed, the less the influence of the sampling rate will be. More importantly, for a relatively low sampling frequency, the limited switching frequency may result in undesirable effects on the input signal or even instability of the closed-loop system.

Alternatively, the idea of discrete-time sliding mode control (DSMC) has been proposed in literature, which is significantly different from its continuous-time counterpart; see [83] for more information. The results presented in e.g. [83] demonstrate that an appropriate choice of sliding surface, used with the *equivalent control*, can ensure a bounded motion about the surface in the presence of bounded matched uncertainty. Notice also that from this viewpoint, the DSMC problem can be seen as a robust optimal control problem and is related to discrete-time Lyapunov min-max problems [83]. The problem is to select, among all possible feedback controllers, the feedback gain that minimizes the worst case effect of the uncertainty on the Lyapunov difference function [83]. Moreover, the discrete-time equivalent control law can be considered as a solution of the discrete-time linear quadratic regulator (LQR) problem under the assumption of *cheap control*; that is, no penalty is assigned to the control effort in the cost function.

In this book, we explain our recent investigations to improve DSMC and adopt this control strategy to different fields.

The first introductory chapter (Chapter 1) discusses the reasons to consider DSMC. Furthermore, for tutorial purposes, a brief review of CSMC is given in the context of a second-order system. Lastly, in this chapter, the well-known regular form-based method for the design of SMC is reviewed in the framework of discrete-time systems.

Chapter 2 first provides an overview of the relevant literature and places the contribution of the book in a proper context. Further in this chapter, two new forms of switching function are proposed which can be more efficient in terms of reducing the ultimate bound on the system state and reducing the chattering created by traditional switching functions. This new switching function basically uses a disturbance estimator which comes from the same idea presented in [133]. The main idea is, with the assumption of continuity of the original continuous-time disturbance signal, to use the previous value of the sampled disturbance for estimating the current one in the control law. However, model uncertainty is not considered in [133]. In Chapter 2, it is also discussed that using the mentioned estimator directly in the controller will increase the order of the system and, in addition, it results in a system involving time-delay. Stability analysis and ultimate boundedness are then investigated for this kind of system. This method greatly reduces the conservatism of the current linear matrix inequality (LMI)-based methods presented in the few existing works that consider the problem of applying DSMC to the systems including unmatched uncertainties. Specifically, this method avoids using inequalities to deal with the uncertain negative signum quadratic terms appearing in the derived Riccati-like inequality, which is not easy to be directly arranged as an LMI problem. Instead, a *lossless* technique is proposed to convert the mentioned inequality to a form that can be easily written as an LMI. These results were previously published in the paper [13].

While Chapter 2 proposes a state feedback DSMC for uncertain discrete-time systems whose whole states' information is available, Chapter 3 proposes an observer-based output feedback DSMC for discrete-time multi-input multi-output (MIMO) systems. This is more practical, as in many real applications, only systems' output is accessible. Furthermore, the disturbance estimator in Chapter 2 has been designed for the cases that the system states are entirely available. By exploiting output information only for discrete-time MIMO systems with unmatched disturbances and without uncertainties, a framework has been proposed in [32]. Chapter 3 uses an integral term of the estimation output error, in addition to the well-known Luenberger observer which observes the system state with a proportional loop, to allow more degrees of freedom. This matter is referred to as *proportional integral observer* (PIO) in the literature [32]. Nevertheless, the underlying system in [32] does not involve unmatched uncertainties, unlike the system considered in this chapter. The proposed *scheme* here extends the problem of utilizing disturbance observer in the output feedback DSMC (ODSMC) to uncertain discrete-time systems using an innovative LMI based framework. Many of the results in Chapter 3 were previously published in the conference paper [11].

The main goal of Chapter 4 is to stabilize a networked control system (NCS) involving consecutive data packet dropout with a sliding mode control strategy that can improve the existing approaches. In doing so, a novel sliding function is introduced by employing the available communicated system states involving packet losses. This is significantly different from the existing DSMC in the literature [101, 33], and it also provides the possibility to directly build the switching component of the DSMC by exploiting only the available system states. The results in Chapter 4 are based on the papers [6, 15].

The DSMC, given for NCSs in Chapter 4, is derived based on two major assumptions:

1. the packet losses occur only in the channel from the sensor to the controller;
2. the system states are entirely available.

However, these assumptions may be unrealistic for many practical problems. Thus Chapter 5 intends to design sliding mode controllers for NCSs involving both measurement and actuation consecutive packet losses (or long-term random delays), which exploit only output information. This ODSMC can distinguish itself from the existing literature on the SMCs applied to the NCSs, in the sense that both the measurement and actuation delays are viewed as the Bernoulli distributed white sequence. The results in Chapter 5 were previously published in the paper [7].

Decentralized SMC has previously been developed in the literature for large-scale interconnected systems [144, 145, 112, 92]. However, distributed SMC has received less attention and hence it requires more investigation. Chapter 6 first explores the problem of designing a sparse DSMC network for a given plant network with arbitrary topology. To do so, this chapter considers a priori the control network topology which is a subset of the underlying dynamics network and provides a methodology to stabilize the underlying dynamics utilizing a (sparse) distributed observer and controller network. We will show that the proposed observer-based DSMC has the ability to cover all the cases such as decentralized, distributed, and sparsely distributed topologies. In Chapter 6, as the second step, we will search for a sparse control/observer network structure with the least possible number of links that can satisfy the given stability condition. To this end, a heuristic iterative algorithm will be proposed, distinguishing itself from a trial-and-error process which requires checking of all the possible structures. These results were previously published in the conference paper [14].

Although the SMC is now a well-known strategy, from the standpoint of constraining the available control action, all the traditional methods considered in the literature have shortcomings. This drawback basically comes from the nature of the SMC design process which contains two separate stages. During the synthesize of the sliding function, there is no sense of the control action level that is required to induce and retain sliding. This issue is more crucial in Chapter 7 when it comes to sparsifying the control network structure, as without limits on the available control actions, it may result in the high level of control efforts that each subsystem's controller requires to apply, which is not a practical case. Chapter 7 develops an approach by which we can deal with an $\mathscr{H}_2$ based optimal structured SMC problem. In this chapter in order to address the problem of designing a sparse SMC controller, a specific form of fictitious system, whose matrices contain the control network structure, is derived. This makes the well-developed weighted ℓ_1 algorithm infeasible to apply to our problem. Alternatively, Chapter 7 proposes a heuristic scheme to obtain the sparse sliding mode controller. The results in Chapter 7 were published in the papers [12, 8].

According to the so-called 1D quasi-sliding mode, SMC design has been extended for 2D systems in the Roesser Model (RM). In addition, the conditions to ensure the remaining horizontal and vertical states in RM on the switching surfaces and also the reaching condition using a 2D Lyapunov function are investigated in [3]. Another strategy to work with 2D systems is to transfer them to a 1D form. Wave advance model (WAM) is a 1D form of 2D systems established in [111]. From the view point of WAM model, 2D systems are considered as advanced waves and consequently the original stationary 2D system is converted to a time-varying 1D system. Moreover, the system matrices are in rectangular form rather than square form. As a result, the major drawback of this 1D form of 2D systems is the varying dimensions of the defined state vectors. This means that the results developed using this framework are most likely computationally unattractive in terms of possible applications. Motivated by this issue and by the use of stacking vectors, a new approach to converting 2D systems to a 1D form is proposed in Chapter 8. Consequently, the states, inputs and outputs of the obtained 1D system are in the vector form, and more importantly their dimensions are invariant. This framework is basically useful for a class of 2D linear systems in which information propagation in one of the two distinct directions only occurs over a finite horizon. This can be the case of a repetitive process [50] or any inherently 2D system, for instance, the Darboux equation [73]. The suggested 1D vectorial form in Chapter 8 unlike the WAM form has invariable dimension and consequently can be converted to *regular form* in SMC. In this chapter, first the Fornasini and Marchesini (FM) model of 2D systems which is a second order recursive form is considered. The results in Chapter 8 for 2D systems were published in the paper [5].

In Chapter 9, first, the controllability analysis of the WAM model of the first FM model is studied, and a necessary condition for the controllability of this 1D model is given. On the other hand, during the procedure of designing the sliding surface in Chapter 8, it is assumed that the obtained 1D system is controllable. But, the controllability of the obtained 1D form and its relation to the original 2D system is an unanswered problem in Chapter 8. Hence, motivated by these issues, in this chapter, we focus on the controllability analysis of the proposed 1D form of the underlying 2D systems. Based on the controllability analysis, a new notion, *directional controllability*, for the underlying 2D systems is introduced and studied. More importantly, a necessary and sufficient condition for the directional controllability of 2D systems is presented in this chapter. The controllability analyses of 2D systems here were published in the papers [9, 10].

Finally, Chapter 10 is devoted to the problem of heart rate regulation during cycle-ergometer exercise using both a non-model-based as well as a model-based control strategy along with a real-time damped parameter estimation scheme. The model-based control strategy is a time-varying integral sliding mode controller. A recursive damped parameter estimation method is also developed, by incorporation of a weighting upon the one-step parameter variation, which in contrast to the conventional parameter estimation schemes can avoid the occurrence of the so-called blowup phenomena. The calculated control signals are transmitted to the subjects employing a synchronized biofeedback mechanism. Indeed, delivering a feedback signal when the pedals are not in a suitable position to efficiently exert force may be ineffective and this may, in turn, lead to the cognitive disengagement of the user from the feedback controller. Chapter 10 examines a novel form of control system which has been designed for this project. The system is called an "actuator-based event-driven control system". The proposed control and estimation scheme were experimentally verified using several healthy male participants and the results demonstrated that the designed scheme is able to regulate the HR of the exercising subjects to a predetermined HR profile preventing overshooting in the HR responses. The results in this chapter are based on the published papers [16, 17, 18, 19, 20, 21].

Ahmadreza Argha
Steven W. Su
Li Li
Hung T. Nguyen
Branko G. Celler

Sydney, Australia

Contributors

Dr. Ahmadreza Argha
University of New South Wales
Sydney, NSW, Australia
Dr. Ahmadreza Argha received B.S. and M.S. degrees in Electrical Engineering from Shiraz University, Iran. He also received a Ph.D. from the University of Technology Sydney (UTS), Australia in 2017. He is currently a Postdoctoral Research Fellow at the University of New South Wales (UNSW), Australia. His research interests include biomedical system modeling and control, robust control, and network systems.

A/Prof. Steven W. Su
University of Technology Sydney
Sydney, NSW, Australia
Associate Professor Steven W. Su received B.S. and M.S. degrees from Harbin Institute of Technology (HIT), China, and earned a Ph.D. degree from the Australian National University (ANU), Australia. He is currently an Associate Professor in the Faculty of Engineering and IT, at the University of Technology Sydney (UTS). His major research interests include biomedical system modeling and control, robust and adaptive control, and rehabilitation engineering.

A/Prof. Li Li
University of Technology Sydney
Sydney, NSW, Australia
Associate Professor Li Li received his B.S. degree from Huazhong University of Science and Technology in 1996, his M.S. degree from Tsinghua University in 1999, and his Ph.D. degree from University of California, Los Angeles in 2005. He joined the University of Technology Sydney (UTS) in 2011 and currently he is an Associate Professor in the Faculty of Engineering and IT. His research interests are control theory and power system control.

Prof. Hung Nguyen
University of Technology Sydney
Sydney, NSW, Australia
Professor Hung T. Nguyen received his B.E. degree with First Class Honours and University Medal in 1976 and earned his Ph.D. degree in 1980 from the University of Newcastle in Australia. He is the Professor of Faculty of Engineering and IT at the University of Technology Sydney (UTS). He has been involved with research in the areas of biomedical engineering, artificial intelligence, neurosciences and ad-

vanced control for more than 20 years. Prof. Nguyen is a Fellow of the Institution of Engineers, Australia; the Australian Computer Society; and the British Computer Society.

Prof. Branko Celler
University of New South Wales
Sydney, NSW, Australia
Professor Branko G. Celler received B.Sc. and B.E. (Hons.) degrees in Electrical Engineering and earned a Ph.D. degree in Biomedical Engineering from the UNSW, Australia. He is currently a Research Professor at the University of NEW South Wales School of Electrical Engineering. His research interests include biomedical instrumentation and systems, noninvasive modeling of cardiovascular performance, and medical informatics. Prof. Celler is a Fellow of IEEE and a Fellow of the Academy of Technological Sciences and Engineering (ATSE).

1

Introduction

CONTENTS

1.1 Continuous-time SMC 2
1.2 Regular form-based DSMC 20
1.3 Summary 22

Abstract– *Why discrete-time sliding mode control?*
While a large number of investigations in the control systems literature focus on the analysis of continuous-time systems, more and more practising control engineers implement the control laws using digital computers. The controllers can either be carried out from continuous-time representations using fast sampling ideas, or the continuous-time controllers can be converted to their discrete-time representations. However, the choice of the high sampling rate, which nearly approximates continuous-time, may not always be possible. Alternatively, discrete-time controllers can be designed directly from a discrete-time representation of the plant. As a result, one thread of the literature develops discrete-time controllers to stabilize discrete-time linear systems.

In this book, our main focus is on the design of a specific control strategy using digital computers. This control strategy referred to as sliding mode control (SMC) has its roots in (continuous-time) relay control. In fact, as the SMC technique relies on an infinite switching frequency of the input signal, it is inherently a continuous-time control strategy. However, this matter can never be met in real applications, especially for discrete-time controllers where the input signal can only be varied at the sampling instances. This fact can limit the switching frequency to the sampling frequency. Nevertheless, in the case that the sampling rate is much faster than the dynamics of the system under control, the influence of the bounded switching frequency will be confined. It is thus a usual approach to design sliding mode controllers in the continuous-time domain, even if the system is computer-aided-controlled [149], regarded as continuous-time sliding mode controller (CSMC), since it is designed according to a continuous-time model of the system, regardless of the sampling issue. However, the effectiveness of the obtained controller will strongly depend on the sampling frequency, i.e. the faster sampling is performed, the less influence of the sampling rate will be. On the other hand, for a relatively low sampling frequency, the limited switching frequency may result in undesirable effects in the input signal or even instability of the closed-loop system.

This book aims to explain our recent research outcomes in the field of discrete-time sliding mode control (DSMC). The discrete-time systems here are assumed to be obtained by exploiting the sample-and-hold method of sampling from continuous-time systems. In what follows, we present a brief introduction to the concept of continuous-time SMC, and the regular form-based method for the design of SMC, albeit in the context of discrete-time systems.

1.1 Continuous-time SMC

While considering practical control problems, a discrepancy may exist between the actual system and the model used to describe the system behavior; i.e. what is the system output with a specific input. Discrepancies can occur due to exogenous disturbances, unmodeled dynamics, etc. Usually, in model-based control design schemes, this (inaccurate) mathematical model is used for the design of a controller. As a result, controllers should be able to provide a desired performance for the closed-loop system in the presence of disturbances/uncertainties. This task is the main target of the so-called robust control methods. Sliding mode control technique is indeed one of the robust control approaches among many methods proposed and considered in control theory.

Consider the following uncertain linear-time-invariant (LTI) continuous-time system:

$$\dot{x}(t) = Ax(t) + B[u(t) + \xi(x,u,t)], \tag{1.1}$$

where $x \in \mathbb{R}^n$ and $u \in \mathbb{R}^m$ are the state vector and control input vector. The unknown signal $\xi(x,u,t) : \mathbb{R}^n \times \mathbb{R}^m \times \mathbb{R}_+ \to \mathbb{R}^m$ denotes the matched uncertainty in (1.1) whose Euclidean norm is bounded by a known function.

Definition 1.1 *Consider the following system*

$$\dot{x}(t) = Ax(t) + Bu(t) + \tilde{B}\tilde{\xi}(x,u,t). \tag{1.2}$$

The uncertainty $\tilde{\xi}$ in (1.2) is said to be (un)matched uncertainty, if the range space of the input matrix B (does not) contains the range space of $\tilde{B}$ [43].

Without loss of generality, assume that the matrix B has full rank and $m \leq n$. For example, consider a double integrator system, i.e. A and B matrices in (1.1) are as follows

$$A = \begin{bmatrix} 0 & 1 \\ 0 & 0 \end{bmatrix}, \; B = \begin{bmatrix} 0 \\ 1 \end{bmatrix}. \tag{1.3}$$

Now let us design a control law for u that asymptotically steers the system states to the origin; i.e. $x = 0$. As the first choice, let us consider $u = Fx$, where $F \in \mathbb{R}^{1\times 2}$ is a feedback gain matrix which can be designed using numerous available

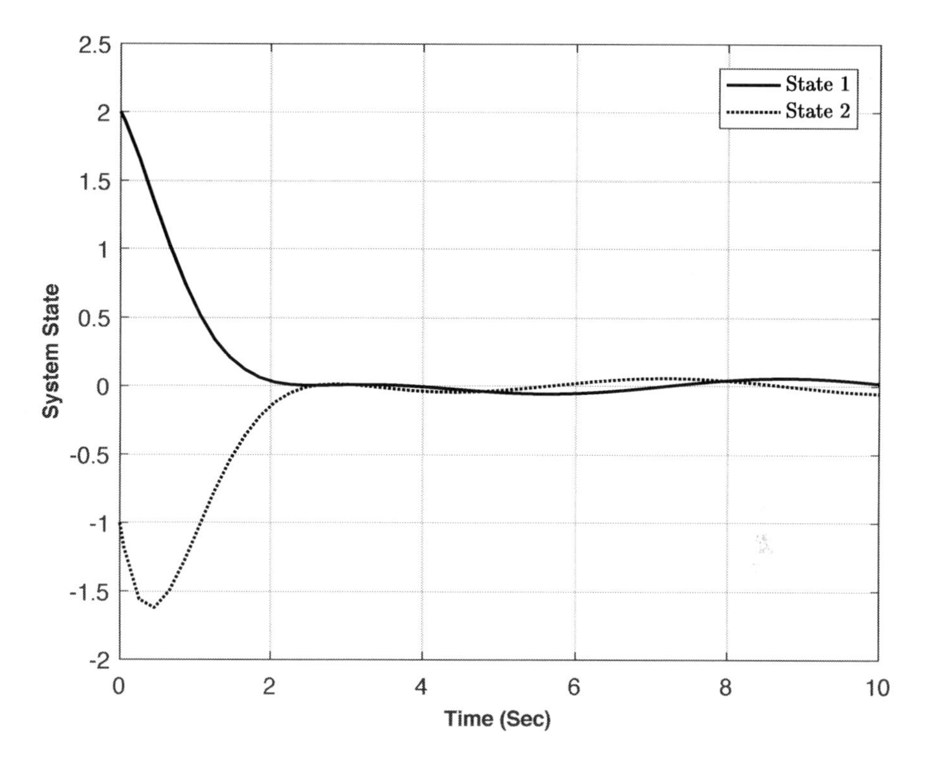

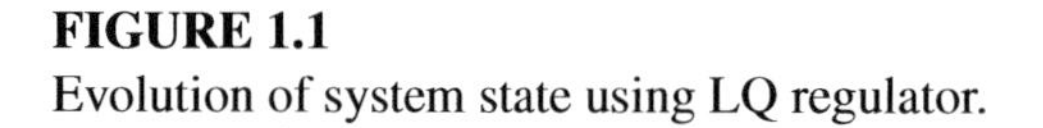

FIGURE 1.1
Evolution of system state using LQ regulator.

approaches. We design F using the linear quadratic regulator (LQR) design approach with $Q = \begin{bmatrix} 10 & 0 \\ 0 & 1 \end{bmatrix}$ and $R = 1$. The obtained gain is $F = \begin{bmatrix} -3.1623 & -2.7064 \end{bmatrix}$, and the poles of the closed-loop system $A + BF$ are located at $-1.3532 \pm 1.1537i$. Fig. 1.1 depicts the evaluation of system states with the proposed LQ regulator when the initial conditions are $x(0) = \begin{bmatrix} 2 & -1 \end{bmatrix}^T$ and $\xi(x,u,t) = 0.2\sin(t)$. As it is evident from Fig. 1.1, this controller cannot asymptotically steer all states to the origin in the presence of ξ. In other words, the LQ regulator can only steer the system states into a region within a bound about $x = 0$. Now, define a new variable σ as

$$\sigma = x_2 - Mx_1, \tag{1.4}$$

where M is a (scalar) design parameter which should be designed such that if $\sigma = 0$ the remaining dynamics are stable. From $\sigma = 0$, we can derive

$$x_2 = Mx_1, \tag{1.5}$$

Substituting (1.5) into $\dot{x}_1 = x_2$, we can obtain $\dot{x}_1 = Mx_1$. This is indeed the dynamics which describes sliding motion. Thus to ensure stability in sliding mode, M

should be a negative scalar. As can be seen from $\dot{x}_1 = Mx_1$, the disturbance ξ has no influence on the sliding mode. From the condition $\dot{\sigma} = \dot{x}_2 - M\dot{x}_1 = 0$, we may obtain

$$\dot{\sigma}(t) = -Mx_2(t) + u(t) + \xi(x,u,t) = 0, \quad \forall t > t_s \tag{1.6}$$

where t_s denotes the time when sliding motion starts. To satisfy $\dot{\sigma} = 0$, a control law can be derived as

$$u_{eq}(t) = Mx_2(t) - \xi(x,u,t). \tag{1.7}$$

This is the so-called equivalent control and is not implementable as ξ is unknown. Indeed, the equivalent control can be regarded as the average control effort required to stay sliding. Now rather than the equivalent control, consider the following control law:

$$u(t) = Mx_2(t) - \rho\,\text{sign}(\sigma(t)). \tag{1.8}$$

It can be shown the above control law can steer σ to zero in finite time if $\rho = \bar{\xi} + \varepsilon$, where $\bar{\xi} > 0$ is a known upper bound on the disturbance ξ, i.e. $\|\xi(x,u,t)\| \leq \bar{\xi}$ and $\varepsilon > 0$ is a small scalar. Consider a candidate Lyapunov function as

$$V = \frac{1}{2}\sigma^2. \tag{1.9}$$

Now,

$$\begin{aligned} \dot{V} &= \sigma\dot{\sigma} = \sigma(Mx_2 - Mx_2 + \xi - \rho\,\text{sign}(\sigma)) \\ &\leq |\sigma|(\bar{\xi} - \rho) = -\varepsilon|\sigma|. \end{aligned} \tag{1.10}$$

This shows the finite-time convergence of the sliding function σ. Note that $\sigma\dot{\sigma} < 0$ is known as reachability condition. Now, we apply the SMC in (1.8), with $M = -1.8875$ and $\rho = 4$, to the system in (1.1) with $\xi(x,u,t) = 0.2\sin(t)$. The results are illustrated in Figs. 1.2-1.6. As it is evident from Figs. 1.2 and 1.3, the SMC in (1.8) ensures the finite-time convergence of the sliding function as well as asymptotic convergence of system states to zero when $\xi \neq 0$. The reaching phase and the sliding phase can be seen in Fig. 1.6. However, as can be seen in Figs. 1.4 and 1.5, due to the practical limitations on the sign function implementation, the so-called chattering phenomenon occurs while using the SMC (1.8). It is worth noting that in some applications such a switching is inherent, e.g. electrical converters. However, broadly speaking, in many other applications the high frequency switching is undesirable [98].

Since the actuator bandwidth is usually limited, an infinite switching frequency is not achievable. Also, the high frequency control signals in real applications may have harmful consequences, e.g. large current peaks in electrical actuators and high wear in mechanical gear boxes. One simple and useful method to make the discontinuous component in (1.8) continuous and smooth is approximating sign$(\cdot)$ by some continuous/smooth function. For example, sigmoid function is a well-known choice [43]:

$$u_n = -\rho\frac{\sigma}{|\sigma| + \varepsilon}, \tag{1.11}$$

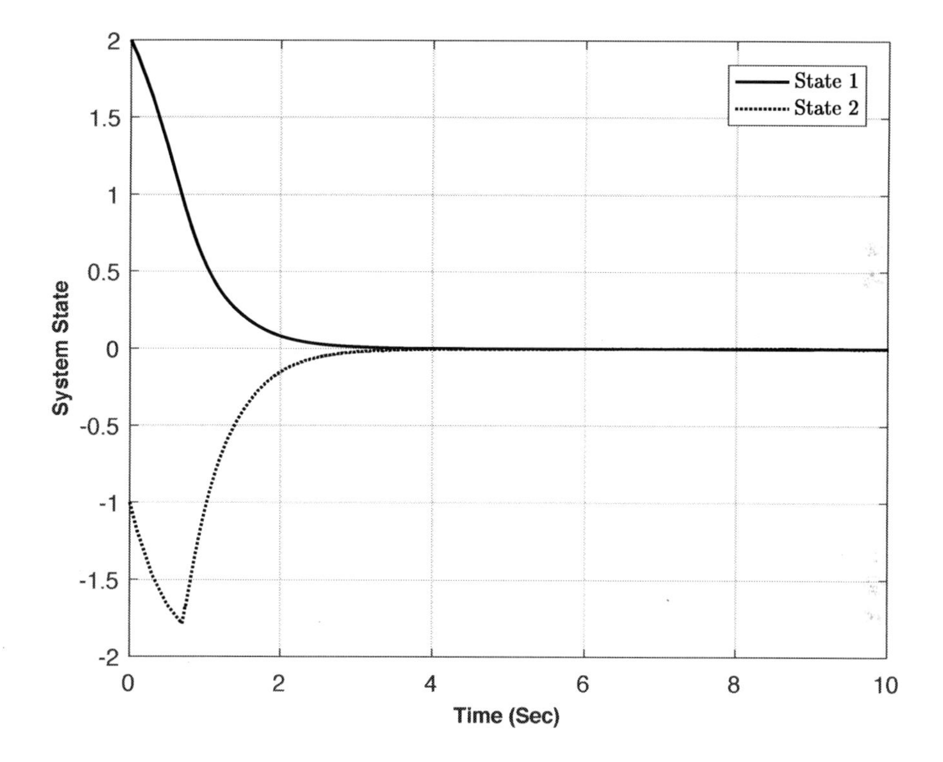

FIGURE 1.2
Evolution of system state obtained by applying SMC in (1.8) with $M = -1.8875$ and $\rho = 4$ to the system 1.1.

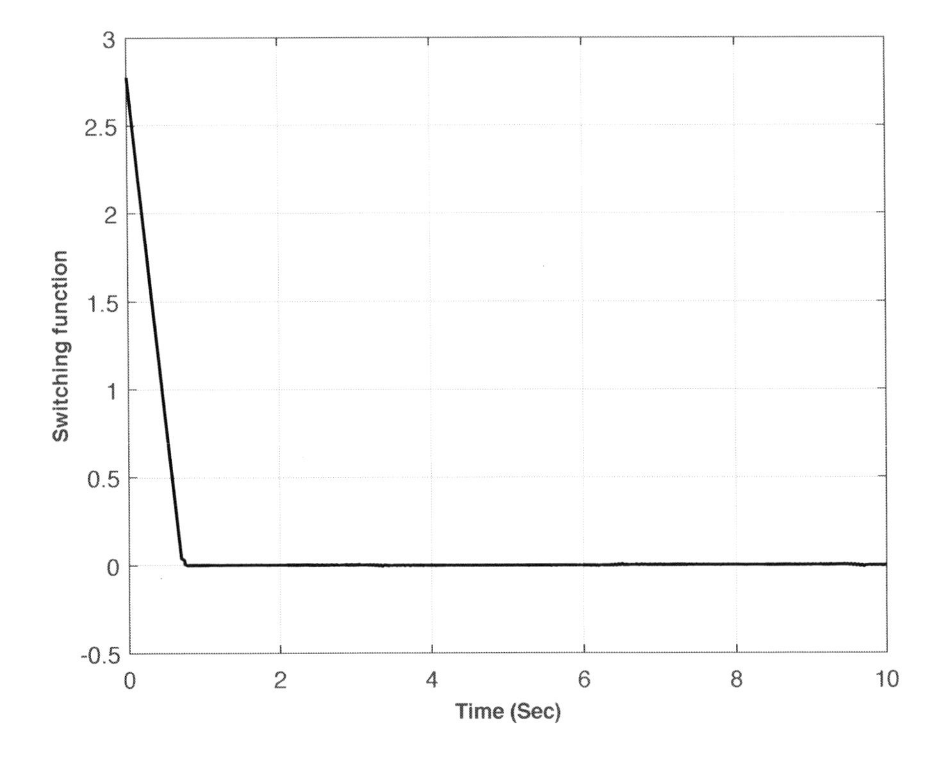

FIGURE 1.3
Evolution of switching function for the system 1.1 using SMC in (1.8) with $M = -1.8875$ and $\rho = 4$.

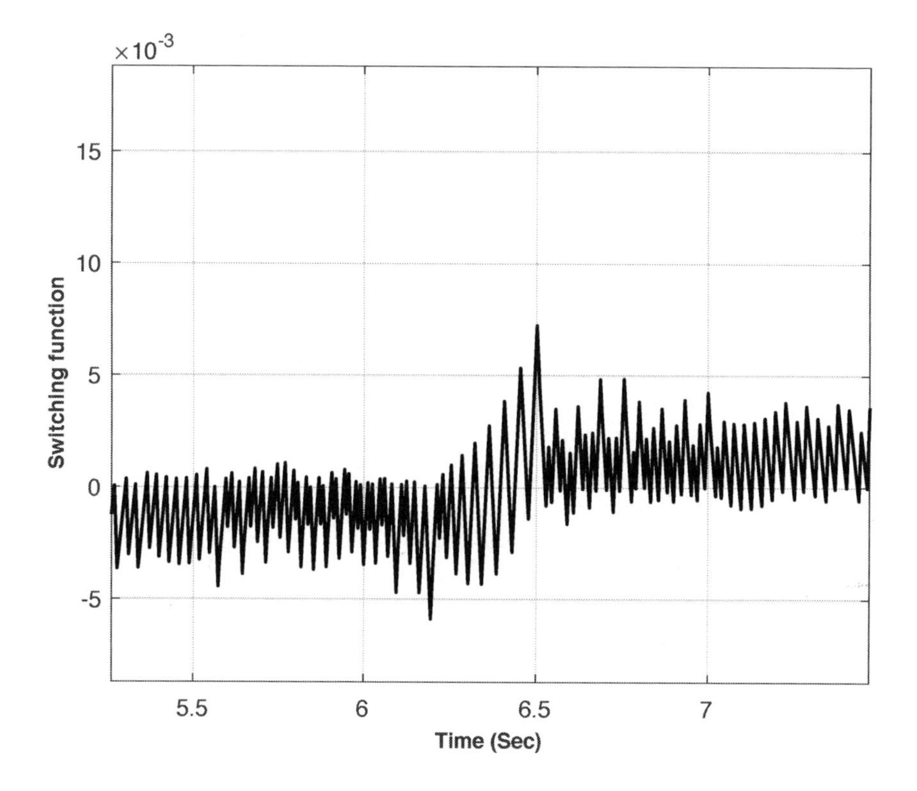

FIGURE 1.4
Evolution of switching function (zoom) for the system 1.1 using SMC in (1.8) with $M = -1.8875$ and $\rho = 4$.

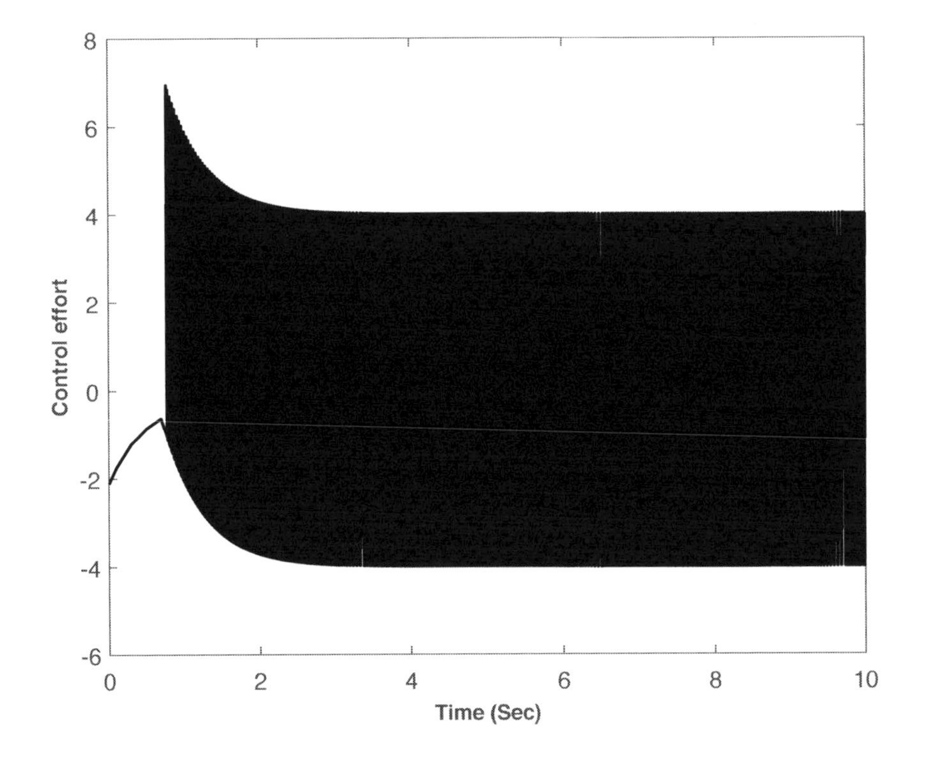

FIGURE 1.5
Control effort with SMC in (1.8) with $M = -1.8875$ and $\rho = 4$.

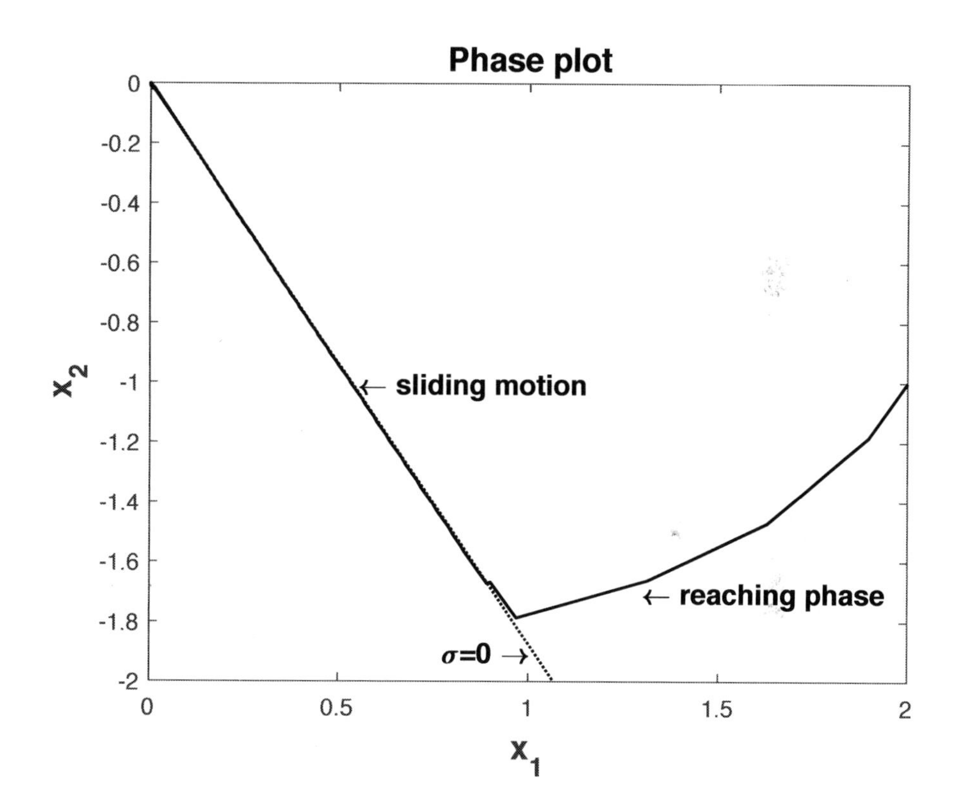

FIGURE 1.6
Phase plot for the system (1.1) using SMC in (1.8) with $M = -1.8875$ and $\rho = 4$.

where $\varepsilon > 0$ is a small scalar and u_n denotes the nonlinear controller part of the SMC (1.8). Note that ε is a design freedom to trade off between having an ideal performance and ensuring a smooth control signal. Let us replace $\rho\text{sign}(\sigma)$ with $\rho\frac{\sigma}{|\sigma|+\varepsilon}$ in (1.8) to yield:

$$u(t) = Mx_2(t) - \rho\frac{\sigma}{|\sigma|+\varepsilon}. \tag{1.12}$$

Applying this new controller, with $M = -1.8875$, $\rho = 4$ and $\varepsilon = 0.1$, to the system (1.1) leads to the results shown in Figs. 1.7-1.9. As it is evident from these results, the controller (1.12) does not lead the switching function σ to converge to the origin in finite-time when $\xi \neq 0$, and further the system states do not converge to zero. However, the sliding variable converges to a bound around $\sigma = 0$ and the system states converge to a region within a bound about $x = 0$. The controller (1.12) is referred to as *quasi sliding mode control* and the boundary region about $\sigma = 0$ described previously is called *quasi sliding mode band*.

Now a more practical controller rather than (1.12) can be proposed as

$$u(t) = Mx_2(t) + \varphi\sigma - \rho\frac{\sigma}{|\sigma|+\varepsilon}, \tag{1.13}$$

where $\varphi < 0$ is a scalar which can be used along with ρ to change the convergence rate of sliding variable to a bound around $\sigma = 0$. Note that by the choice of sliding surface (1.4), it follows from (1.6) that

$$\dot{\sigma}(t) = \varphi\sigma - \rho\frac{\sigma}{|\sigma|+\varepsilon} + \xi(x,u,t). \tag{1.14}$$

Let us analyze the reachability of the bound $|\sigma| \geq \delta$, where $\delta = \frac{\bar{\xi}\varepsilon}{\rho-\bar{\xi}}$, with the new controller (1.13). We can consider the reachability condition

$$\sigma\dot{\sigma} \leq -\eta|\sigma|, \tag{1.15}$$

where $\eta > 0$ is a small scalar. Now it follows from (1.15) and (1.14) that

$$\begin{aligned} \sigma\dot{\sigma} &= \sigma\left(\varphi\sigma - \rho\frac{\sigma}{|\sigma|+\varepsilon} + \xi\right) \\ &\leq |\sigma|\left(\varphi|\sigma| - \rho\frac{|\sigma|}{|\sigma|+\varepsilon} + \bar{\xi}\right) \\ &\leq |\sigma|\left(\bar{\xi} - \rho\frac{|\sigma|}{|\sigma|+\varepsilon}\right) \\ &\leq -\eta|\sigma|. \end{aligned} \tag{1.16}$$

The above inequality leads us to

$$|\sigma| \geq \frac{(\eta+\bar{\xi})\varepsilon}{\rho-\eta-\bar{\xi}}. \tag{1.17}$$

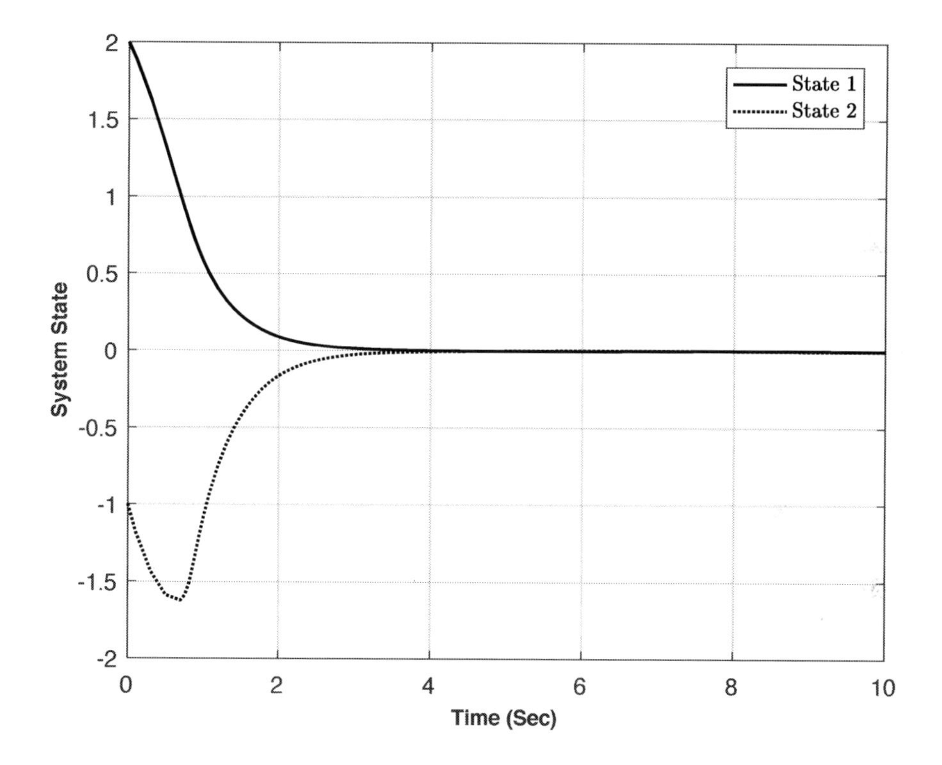

FIGURE 1.7
Evolution of system state obtained by applying SMC in (1.12) with $M = -1.8875$, $\rho = 4$ and $\varepsilon = 0.1$ to the system (1.1).

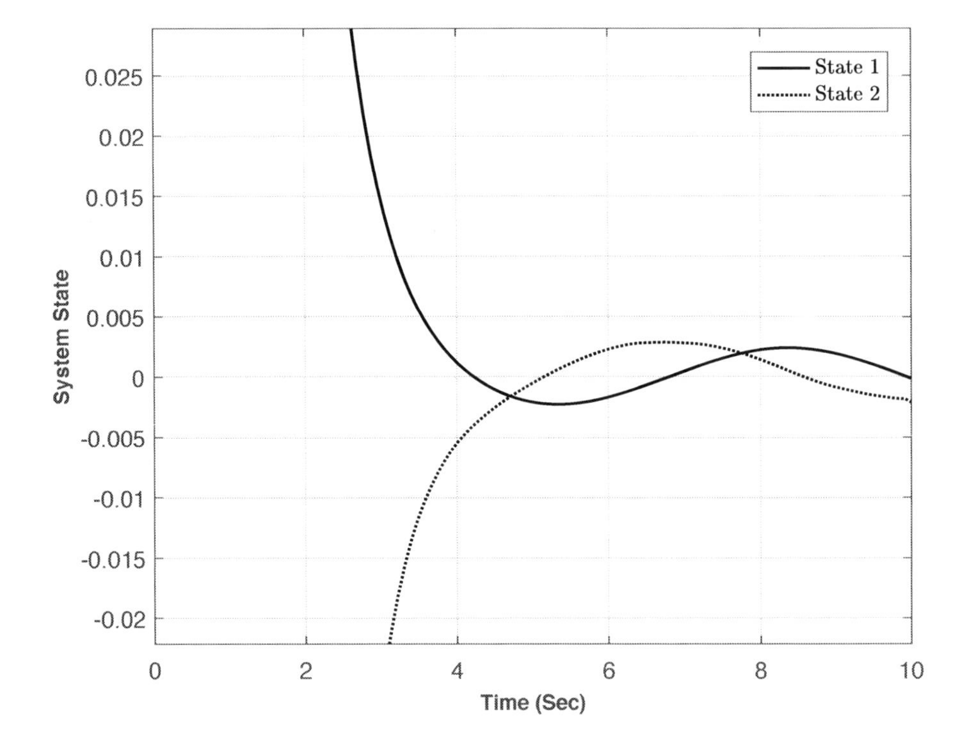

FIGURE 1.8
Evolution of system state (zoom) obtained by applying SMC in (1.12) with $M = -1.8875$, $\rho = 4$ and $\varepsilon = 0.1$ to the system (1.1).

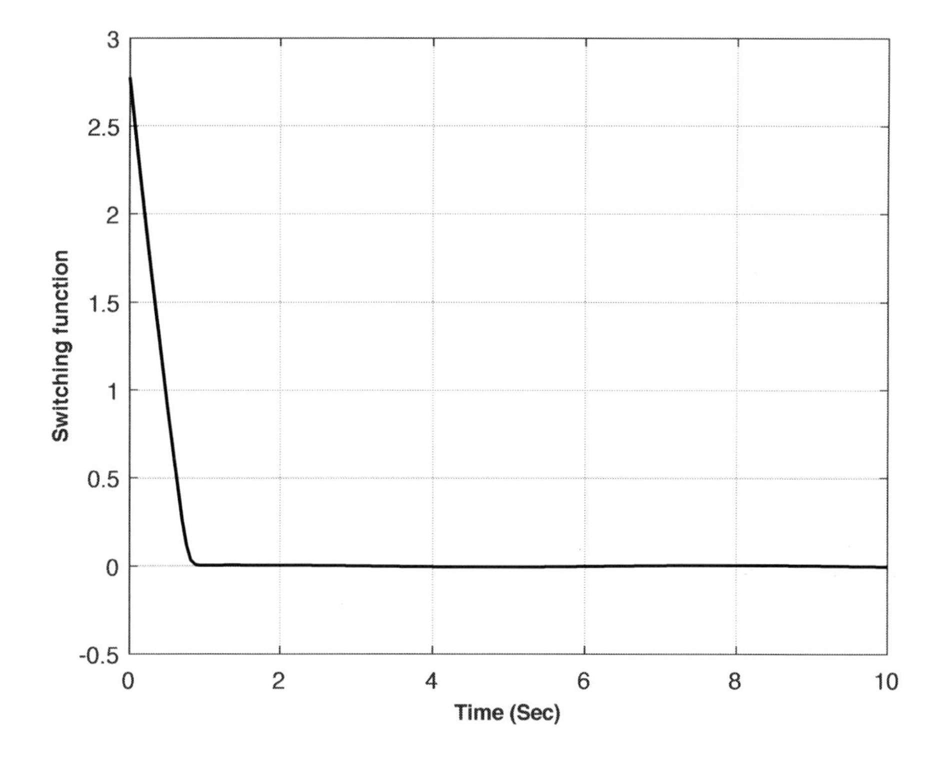

FIGURE 1.9
Evolution of switching function for the system (1.1) using SMC in (1.12) with $M = -1.8875$, $\rho = 4$ and $\varepsilon = 0.1$.

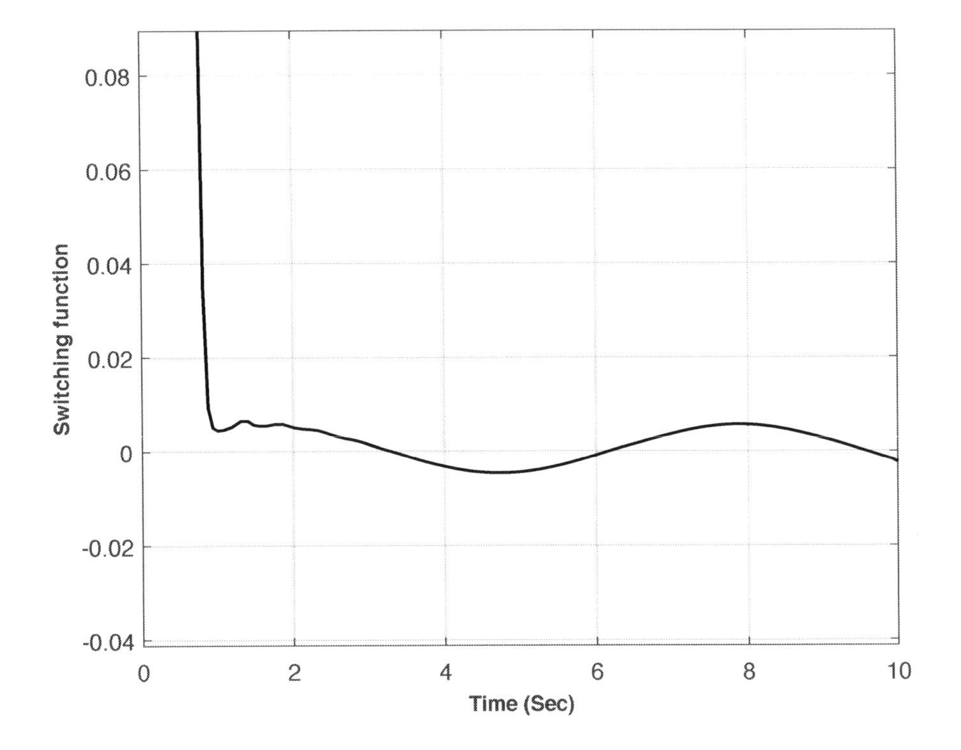

FIGURE 1.10
Evolution of switching function (zoom) for the system (1.1) using SMC in (1.12) with $M = -1.8875$, $\rho = 4$ and $\varepsilon = 0.1$.

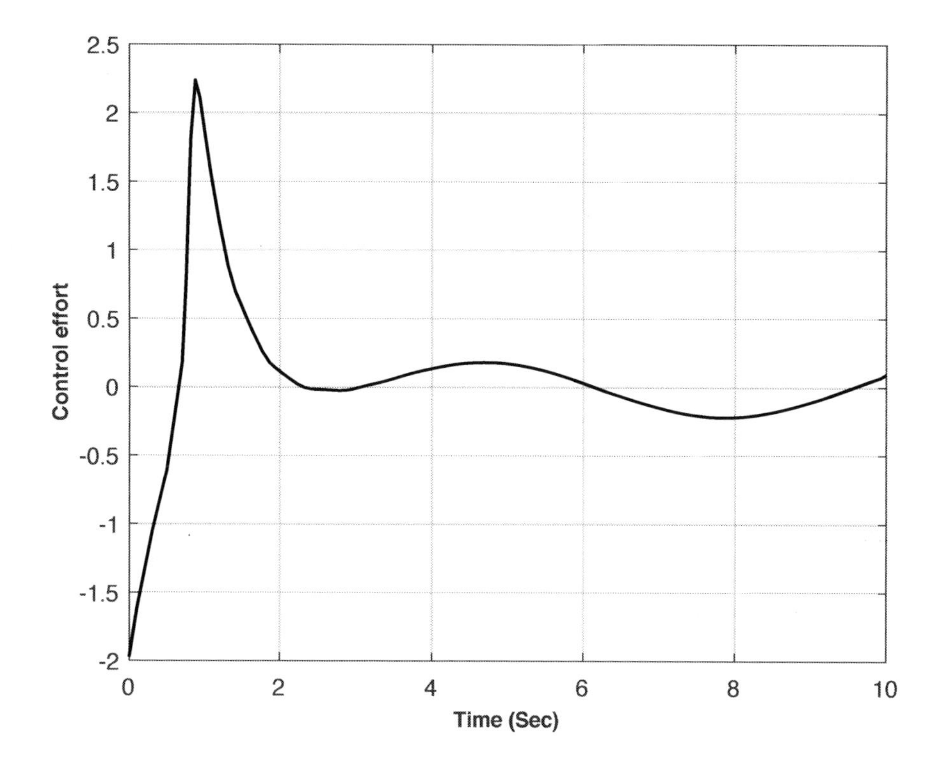

FIGURE 1.11
Control effort with SMC in (1.12) with $M = -1.8875$, $\rho = 4$ and $\varepsilon = 0.1$.

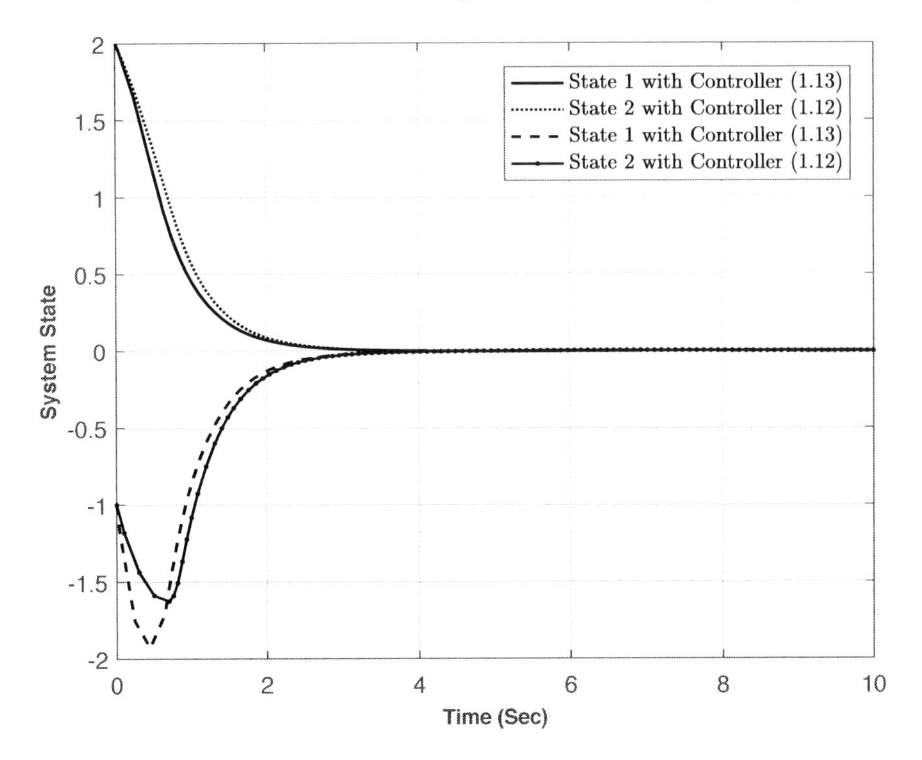

FIGURE 1.12
Evolution of system state obtained by applying SMC in (1.13) and (1.12) with $M = -1.8875$, $\rho = 4$, $\varepsilon = 0.1$ and $\varphi = -1$ to the system (1.1).

By taking the scalar $\eta > 0$ very small, the above given bound on $|\sigma|$ reduces to

$$|\sigma| \geq \frac{\bar{\xi}\varepsilon}{\rho - \bar{\xi}}. \tag{1.18}$$

In summary, if $|\sigma| \geq \frac{\bar{\xi}\varepsilon}{\rho-\bar{\xi}}$, the controller (1.13) will force the system states into the quasi SMC band $\delta = \frac{\bar{\xi}\varepsilon}{\rho-\bar{\xi}}$.

With the same choice of $M = -1.8875$, as used previously, and letting $\varphi = -1$, $\rho = 4$ and $\varepsilon = 0.1$, we apply the controller (1.13) to the system (1.1) and the obtained results are shown in Figs. 1.12-1.15.

As it is evident from Figs. 1.12-1.15 the new parameter φ in the controller can be used, along with the parameter ρ, to set the convergence rate to the sliding surface (or indeed into the quasi sliding band). Quasi sliding is obtained in 0.64 s with the controller (1.13), while it takes around 0.94 s for the controller (1.12) to drive the sliding variable to the quasi sliding band.

The double integrator system considered here is a single-input system, and the

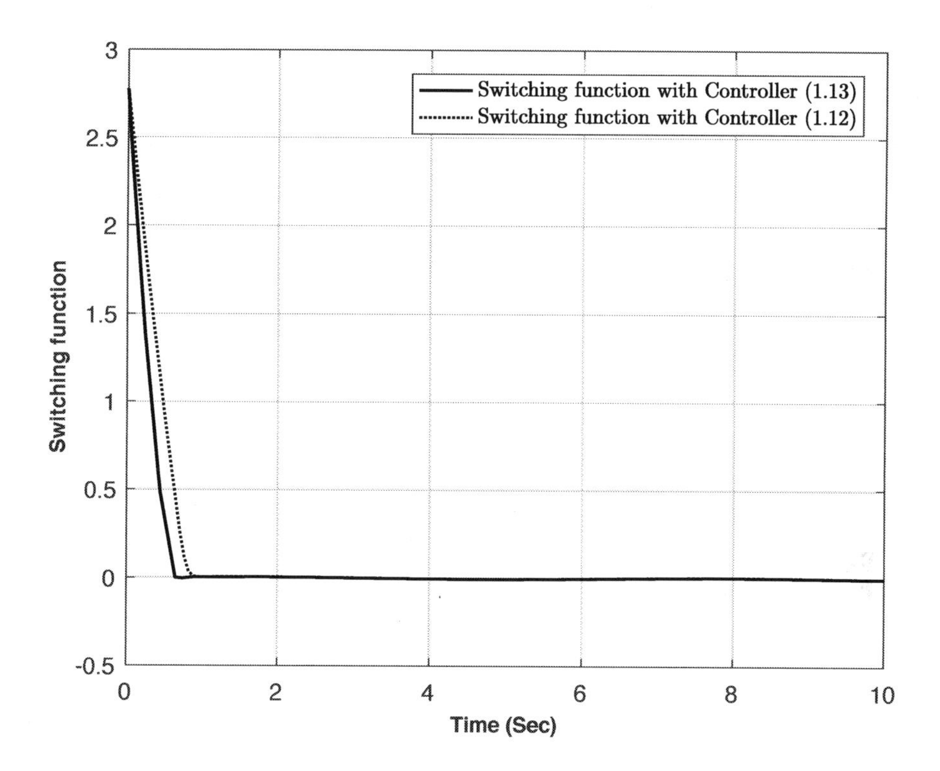

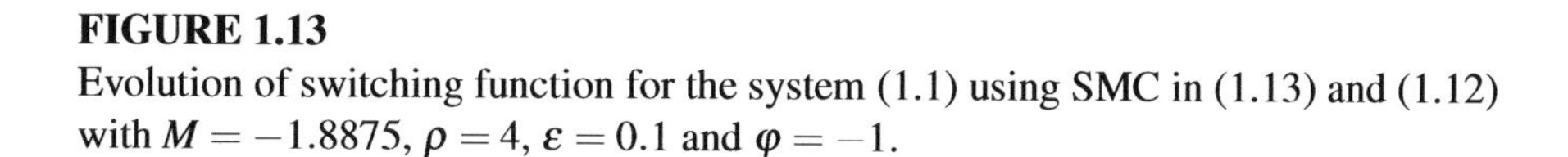

FIGURE 1.13
Evolution of switching function for the system (1.1) using SMC in (1.13) and (1.12) with $M = -1.8875$, $\rho = 4$, $\varepsilon = 0.1$ and $\varphi = -1$.

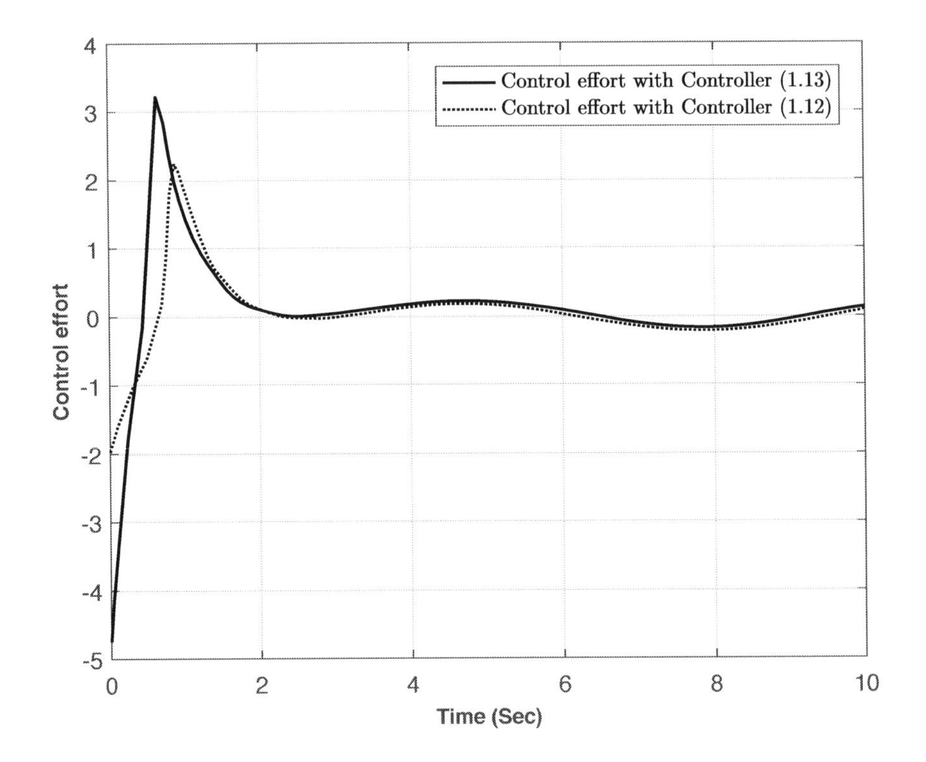

FIGURE 1.14
Control effort with SMC in (1.13) and (1.12) with $M = -1.8875$, $\rho = 4$, $\varepsilon = 0.1$ and $\varphi = -1$.

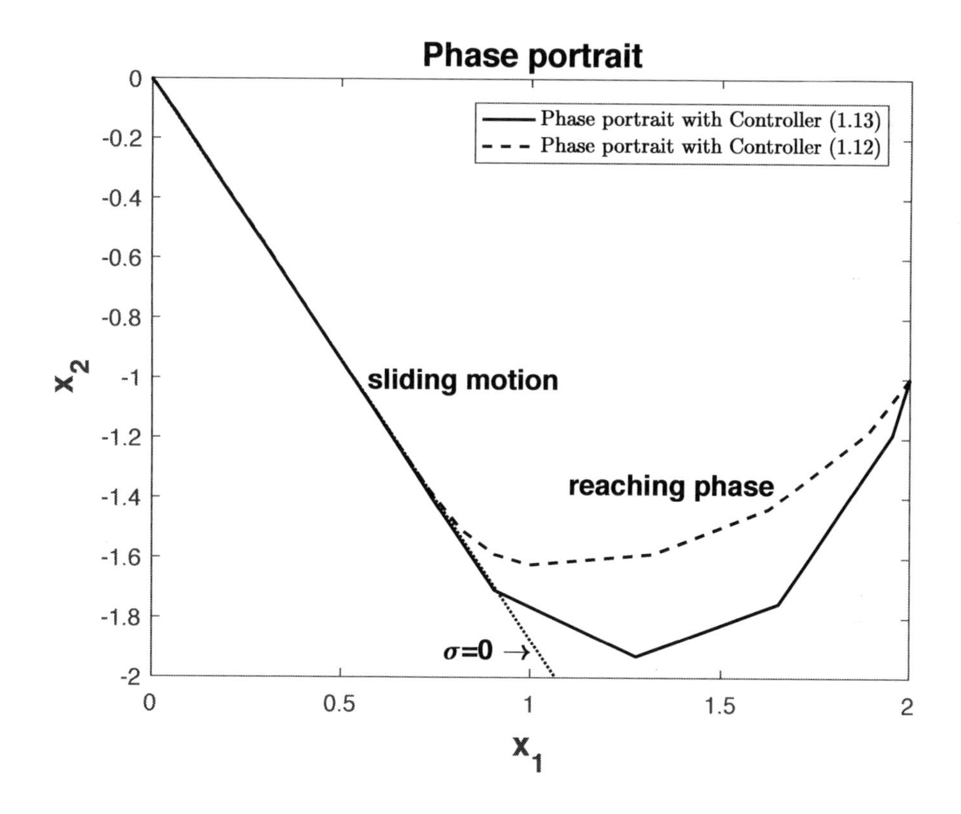

FIGURE 1.15
Phase portrait for the system (1.1) using SMC in (1.13) and (1.12) with $M = -1.8875$, $\rho = 4$, $\varepsilon = 0.1$ and $\varphi = -1$.

sliding mode controller design given above is not extendable to multi-input systems. For multi-input systems, SMC design can be carried out in different ways including the *regular form-based* method. While we give a brief introduction to this method in this chapter, albeit in the context of discrete-time systems, it should be emphasized that the linear matrix inequality (LMI)-based schemes given in the remainder of this book for the design of DSMC is not basically based on the regular form-based SMC design. We only use the regular form-based approach in the design of DSMC for two-dimensional systems in Chapter 8.

1.2 Regular form-based DSMC

Similar to CSMC, the design procedure of the DSMC for stabilizing problems is split into two steps:

1. Design a sliding surface which lead to stable internal dynamics during sliding.
2. Create a control law which drives the closed-loop system into the sliding surface and forces the system trajectories to stay on or at least as close as possible to the surface.

A number of different methods for designing the sliding surface are considered in the literature [43]. In this section, we give a brief introduction to the well-known regular form based approach [43].

Consider the following uncertain linear discrete-time system,

$$x(k+1) = Ax(k) + Bu(k) + f(k), \tag{1.19}$$

where $x(k) \in \mathbb{R}^n$ and $u(k) \in \mathbb{R}^m$. Generally, it is assumed that $B \in \mathbb{R}^{n\times m}$ and $m \leq n$. Besides, $\text{rank}(B) = m$ (matrix B has full column rank) and it is assumed that the pair (A,B) is controllable. Also, $f(k) \in \mathbb{R}^n$ denotes the uncertainty. We consider the following general uncertainty

$$f(k) = \Delta x(k) + d_k, \tag{1.20}$$

where Δ shows the unknown uncertainty with the bound $||\Delta|| < \alpha_0$ ($||.||$ the induced Euclidean or induced spectral norm). Moreover, the term $d_k \in \mathbb{R}^n$, indicates the external disturbance and it is assumed that $||d_k|| < \beta_0$, where β_0 is a known positive constant. As a result, we can write

$$||f(k)|| < \alpha_0 \, ||x(k)|| + \beta_0. \tag{1.21}$$

Since $\text{rank}(B) = m$, there exists an orthogonal matrix $T_r \in \mathbb{R}^{n\times n}$ such that

$$T_r B = \begin{bmatrix} 0_{(n-m)\times m} \\ \bar{B}_2 \end{bmatrix}, \tag{1.22}$$

where the matrix $\bar{B}_2 \in \mathbb{R}^{m\times m}$ is nonsingular [127]. After the coordinate transformation, the system (1.19) is converted to

$$\begin{bmatrix} \bar{x}_1(k+1) \\ \bar{x}_2(k+1) \end{bmatrix} = \begin{bmatrix} \bar{A}_{11} & \bar{A}_{12} \\ \bar{A}_{21} & \bar{A}_{22} \end{bmatrix} \begin{bmatrix} \bar{x}_1(k) \\ \bar{x}_2(k) \end{bmatrix} + \begin{bmatrix} 0_{(n-m)\times m} \\ \bar{B}_2 \end{bmatrix} u(k) + T_r f(k). \tag{1.23}$$

Now, the sliding surface is introduced as

$$\sigma_x(k) = \bar{S}\bar{x}(k) = \bar{S}_1\bar{x}_1(k) + \bar{S}_2\bar{x}_2(k), \tag{1.24}$$

where $\bar{S}_1 \in \mathbb{R}^{m\times(n-m)}$ and $\bar{S}_2 \in \mathbb{R}^{m\times m}$ are the design parameters which determine the sliding surface and should be chosen such that, in the case that $\sigma_x(k) = 0$, all remaining dynamics are stable. During ideal sliding on the surface, $\sigma_x(k) = 0$ for all $k \geq k_s$, where k_s is the time when sliding starts, therefore

$$\bar{x}_2(k) = -\bar{S}_2^{-1}\bar{S}_1\bar{x}_1(k). \tag{1.25}$$

Substituting the equation (1.25) into the equation (1.23) and ignoring the uncertainty $f(k)$ leads to

$$\bar{x}_1(k+1) = (\bar{A}_{11} - \bar{A}_{12}\bar{S}_2^{-1}\bar{S}_1)\bar{x}_1(k). \tag{1.26}$$

Hence, stability in the sliding mode is satisfied when all eigenvalues of the matrix $(\bar{A}_{11} - \bar{A}_{12}\bar{S}_2^{-1}\bar{S}_1)$ are located inside the unit circle. The sliding surface in the original coordinate can be found by $\sigma_x(k) = Sx(k)$, where

$$S = [\bar{S}_1 \ \ \bar{S}_2]T_r. \tag{1.27}$$

Define T_s as

$$T_s = \begin{bmatrix} I_{(n-m)} & 0_{(n-m)\times m} \\ \bar{S}_1 & \bar{S}_2 \end{bmatrix}, \tag{1.28}$$

and in the new coordinate $T_s\bar{x} \longmapsto \tilde{x}$, we have

$$\tilde{x}(k+1) = \begin{bmatrix} \bar{x}_1(k+1) \\ \sigma_x(k+1) \end{bmatrix} = \begin{bmatrix} \tilde{A}_{11} & \tilde{A}_{12} \\ \tilde{A}_{21} & \tilde{A}_{22} \end{bmatrix} \begin{bmatrix} \bar{x}_1(k) \\ \sigma_x(k) \end{bmatrix} + \begin{bmatrix} 0_{(n-m)\times m} \\ \bar{S}_2\bar{B}_2 \end{bmatrix} u(k) + T_sT_r f(k). \tag{1.29}$$

Now, let

$$u(k) = u_l(k) + u_n(k), \tag{1.30}$$

where u_l denotes the linear controller and u_n is the nonlinear component of the DSMC. While we let $u_n = 0$ here, different choices for nonlinear controller in (1.30) will be proposed and discussed later in Chapter 2. Now, consider the following well-known linear sliding control law:

$$u(k) = \left(\bar{S}_2\bar{B}_2\right)^{-1}\left[\left(\Phi - \tilde{A}_{22}\right)\sigma_x(k) - \tilde{A}_{21}\bar{x}_1(k)\right], \tag{1.31}$$

where $\Phi \in \mathbb{R}^{m\times m}$ is a diagonal matrix whose diagonal elements, ϕ_r, $r = 1,\ldots,m$, satisfy $0 \leq \phi_r < 1$. Thus, with the control law (1.31) the closed-loop system is

$$\tilde{x}(k+1) = \begin{bmatrix} \bar{x}_1(k+1) \\ \sigma_x(k+1) \end{bmatrix} = \begin{bmatrix} \tilde{A}_{11} & \tilde{A}_{12} \\ 0 & \Phi \end{bmatrix} \begin{bmatrix} \bar{x}_1(k) \\ \sigma_x(k) \end{bmatrix} + \begin{bmatrix} \tilde{f}_1(k) \\ \tilde{f}_2(k) \end{bmatrix}. \tag{1.32}$$

$$\tilde{x}(k+1) = \tilde{A}_{cl}\tilde{x}(k) + \tilde{f}(k), \tag{1.33}$$

where $\tilde{f} = T_s T_r f$ and

$$\tilde{A}_{cl} = \begin{bmatrix} \tilde{A}_{11} & \tilde{A}_{12} \\ 0 & \Phi \end{bmatrix}. \tag{1.34}$$

Here $\tilde{A}_{cl}$ denotes the stable matrix in the new coordinates obtained by applying the controller in (1.31). The poles of the closed-loop system in the original coordinates (i.e. A_{cl}), when $\Delta = 0$, are given by

$$\lambda(A_{cl}) = \lambda(\tilde{A}_{11}) \cup \lambda(\Phi), \tag{1.35}$$

where $\lambda(\cdot)$ denotes eigenvalues of a matrix. Clearly, the eigenvalues of Φ are stable (by the design choice). In addition, it can be easily proved that $\tilde{A}_{11} = \bar{A}_{11} - \bar{A}_{12}\bar{S}_2^{-1}\bar{S}_1$ which is designed to be a stable matrix by (1.26). Consequently, the system (1.32) is stabilized with the control law (1.31), in the absence of system state uncertainty; i.e. $\Delta = 0$. It should also be mentioned that instead of the control law (1.31), another direct method is possible to be employed in order to obtain the sliding control law [98]. Assume that matrices $\bar{S}_1$ and $\bar{S}_2$ have been designed such that the reduced order system (1.26) is stable. Assume during reaching phase we have

$$\Phi\sigma_x(k) = Sx(k+1). \tag{1.36}$$

Inserting the equation (1.1) in (1.36) leads to

$$\Phi\sigma_x(k) = S[Ax(k) + Bu(k) + f(k)]. \tag{1.37}$$

Therefore, by neglecting the unknown $f(k)$, the control law can be defined to be

$$u(k) = -(SB)^{-1}(SA - \Phi S)x(k). \tag{1.38}$$

This control law is called direct control law which can be obtained directly after computing the sliding matrix S. For the design of $M = -\bar{S}_2^{-1}\bar{S}_1$ in the CSMC, when $\Delta = 0$ and $d_k = B\xi(x,u,t)$, ($\xi \in \mathbb{R}^m$), several methods have been proposed in the literature, including LMI methods, pole placement, LQR-design [43]. However, most of these methods are not applicable to the case with $\Delta \neq 0$. However, this problem was considered in the related literature and several approaches have been proposed for the design of the sliding surface and thereby sliding mode control for the systems with this kind of uncertainties. In the next chapter, we propose an LMI-based approach which uses a loss-less technique to derive the LMI stability condition which is necessary to design the sliding surface.

1.3 Summary

This chapter gave a brief introduction to the basic concepts of continuous-time sliding mode control and discussed the reasons for studying discrete-time sliding mode control. A simple second order example was brought to this chapter to make it easier to discuss different aspects of SMC such as chattering phenomena and the quasi sliding mode concept. A method to overcome the chattering issue was explained in this chapter, i.e. replacing the discontinuous control part by a continuous approximation.

Part I

LMI-Based Discrete-Time Sliding Mode Control

2

LMI-based state feedback discrete-time SMC for uncertain systems

CONTENTS

2.1 Introduction 27
2.2 Problem Formulation 29
2.3 Design of Discrete-time SMC 31
2.3.1 Variable structure discontinuous control considerations 32
2.3.1.1 Using upper and lower bounds of disturbance in the controller: $\mathscr{C}_1$ 33
2.3.1.2 Using $f_i^+ \pm \frac{f_i^-}{2}$: $\mathscr{C}_2$ 34
2.3.2 Design of a robust sliding surface 34
2.3.3 Characterizing system state boundedness 37
2.4 Disturbance Estimate in Control Law 38
2.5 Simulation Results 41
2.6 Conclusions 48

Abstract– In this chapter, a robust LMI-based method is developed for the design of discrete-time sliding mode control (DSMC) for uncertain systems. The proposed robust DSMC can be applied to unstable systems, and also there is no need to stabilize the underlying system first. It is also shown, in this chapter, that with the assumption of smoothness of the external disturbances, a different form of switching element in the controller can outperform the so-called linear controller in terms of the thickness of the boundary layer around the sliding function and the ultimate bound on the system state. Furthermore, the idea of disturbance estimation is extended to uncertain discrete-time systems.

2.1 Introduction

In continuous-time sliding mode control, to achieve ideal sliding mode, in general the control signal must switch at infinite frequency [83]. However, since in digital control strategies, the control signal is held constant during the sampling period, it is normally not possible to achieve ideal sliding. Hence, in uncertain discrete-time systems it is not possible to ensure that the system state remains certainly on a surface within the state space and consequently the DSMC problem is fundamentally different to its continuous-time counterpart [83]. In terms of DSMC, state trajectories will move within a vicinity of the predetermined sliding surface referred to as a quasi-sliding mode band [51].

Although the early works on the DSMC aimed at establishing a discrete-time counterpart to the continuous-time reachability condition [51, 82, 117], it has been shown that DSMC does not necessarily require the use of a variable structure discontinuous control strategy (VSDCS) [67, 127, 98]. References [67, 127] showed that the DSMC without VSDCS can ensure that the state trajectories stay within a neighborhood of the sliding surface in the presence of bounded matched uncertainty. The obtained control law is called *linear control law*. Moreover, according to the results presented in [67, 127], the use of a switching function in the control law may not necessarily improve the performance. Note that, obviously, the DSMC problem using only the linear control law can be regarded as a *robust optimal control problem* and it will be equivalent to discrete-time Lyapunov min-max problems [93] or discrete-time Riccati min-max problems [38]. Nevertheless, some papers in the literature have claimed a better performance thanks to the use of discontinuous components [98]. Indeed, these papers assume that either the sampling rate of the system is very high compared with the maximum frequency component of the exogenous disturbance or the exogenous disturbance is slow (smooth and bounded). With either of these assumptions, the closed-loop system would behave more or less as a continuous-time system [98] and hence, using a discontinuous component in the controller may improve the performance. In this chapter, two new forms of switching function are proposed which can be more efficient in terms of reducing the ultimate bound on the system state and reducing the chattering created by traditional switching functions. These new switching functions, basically, use a disturbance estimator which comes from the same idea presented in [133]. The idea of using a disturbance observer for the DSMC was first presented in [133] and followed by e.g. [89, 96]. The main idea is, with the assumption of continuity of the original continuous-time disturbance signal, to use the previous value of the sampled disturbance for estimating the current one in the control law. However, model uncertainty is not considered in [133]. In this chapter, it is also discussed that using the mentioned estimator directly in the controller will increase the order of the system and, in addition, it results in a system involving time-delay. Stability analysis and ultimate boundedness are then investigated for this kind of system.

It is worth mentioning that a novel implicit Euler numerical scheme has recently been proposed in [2, 66] that can avoid numerical chattering, by using explicit (forward) methods of discretization. However, chattering appears again in the presence of disturbances. The basic idea is to implement the discontinuous input of the DSMC in an implicit form, while keeping its causality (i.e. the controller is non-anticipative). Then this input has to be computed at each sampling time as the solution to a generalized, set-valued equation, which takes the form of a simple projection on an interval in the simplest cases. It should be also noted that implicit methods require an extra computation, and they can be much harder to implement.

Also, note that the problem of designing the DSMC is mainly considered for the systems with matched uncertainty and/or external disturbance [83]. This chapter greatly reduces the conservatism of the current LMI-based methods presented in the few existing works that consider the problem of applying DSMC to the systems including unmatched uncertainties. Specifically, this chapter avoids using inequalities to deal with the uncertain negative signum quadratic terms appeared in the derived Riccati-like inequality, which is not easy to be directly arranged as an LMI problem. Instead, a *lossless* technique is proposed to convert the mentioned inequality to a form that can be easily written as an LMI. This technique can extremely widen the feasible region of the derived LMI condition obtained for the design of robust sliding surface, and hence, the applicability region of our DSMC compared to the existing literature for the DSMC, e.g. see [101, 72]. In brief, the proposed DSMC is a unified framework for general discrete-time LTI systems. This is significantly different from methods whose application is limited to the stable systems, *cf.* [101], and also the methods which need to pre-stabilize the system, *cf.* [102].

The rest of this chapter is organized as follows: Section 2.2 describes the problem formulation. In Section 2.3, the proposed method to design the sliding surface is given. Section 2.4 explains a more practical DSMC for the systems including uncertainty and disturbance. Effectiveness of the proposed DSMC is studied by numerical examples in Section 2.5. Finally, Section 2.6 concludes this chapter.

2.2 Problem Formulation

Consider the following uncertain linear discrete-time system,

$$x(k+1) = [A + \Delta A(k)]x(k) + B[u(k) + f(k)], \tag{2.1}$$

where $x(k) \in \mathbb{R}^n$ and $u(k) \in \mathbb{R}^m$. Without loss of generality, it is assumed that $B \in \mathbb{R}^{n\times m}$ and $m \leq n$. Besides, $\text{rank}(B) = m$ (matrix B has full column rank) and it is assumed that the pair (A,B) is stabilizable. The uncertain matrix $\Delta A(k)$ has the form of:

$$\Delta A(k) = MR(k)N, \tag{2.2}$$

where the matrices M and N are known and $R(k)$ is an unknown matrix satisfying $R^T(k)R(k) \leq I, \forall k \geq 0$; $f(k)$ denotes the external disturbance with known bound, $\|f(k)\| \leq \bar{f}$, where $\bar{f} > 0$. In the rest of this chapter, for simplicity, ΔA_k and ΔA_{k-1} will be used instead of $\Delta A(k)$ and $\Delta A(k-1)$, respectively.

The following lemmas are useful in the sequel.

Lemma 2.1 ([109]) *Let E, $F(k)$ and G be real matrices of appropriate dimensions with $F^T(k)F(k) \leq I, \forall k \geq 0$, then, for any scalar $\varepsilon > 0$, we have*

$$EF(k)G + G^T F^T(k)E^T \leq \varepsilon EE^T + \varepsilon^{-1}G^T G.$$

Corollary 2.1 *Let* $E = \begin{bmatrix} E_1 & E_2 \end{bmatrix}$, $\Delta_1(k) = \begin{bmatrix} F(k) & 0 \\ 0 & F(k-1) \end{bmatrix}$ *and* $H = \begin{bmatrix} H_1 \\ H_2 \end{bmatrix}$ *be real matrices of appropriate dimensions with* $F(k) \in \mathfrak{F}$, $\forall k \geq 0$, *where* $\mathfrak{F} = \{X : X^T X \leq I\}$, *then, for any scalars* $\varepsilon_i > 0$, $i = 1, 2$, *we have*

$$E\Delta_1(k)H + H^T \Delta_1^T(k)E^T \leq E \begin{bmatrix} \varepsilon_1 I & 0 \\ 0 & \varepsilon_2 I \end{bmatrix} E^T + H^T \begin{bmatrix} \varepsilon_1^{-1} I & 0 \\ 0 & \varepsilon_2^{-1} I \end{bmatrix} H.$$

This corollary is a generalization of Lemma 2.1 and can be proved straightforwardly.

Lemma 2.2 *Let* $\tilde{E}$ *and* $\tilde{H}$ *be real matrices of appropriate dimensions, then, for any matrix* $\Pi > 0$, *we have*

$$\tilde{E}^T \tilde{H} + \tilde{H}^T \tilde{E} \leq \tilde{E}^T \Pi \tilde{E} + \tilde{H}^T \Pi^{-1} \tilde{H}.$$

Proof 1 *Note that* $\Pi = \Pi_1 \Pi_1^T > 0$, *where* Π_1 *is an invertible matrix. Then it can easily be proved by*

$$[\tilde{E}^T \Pi_1 - \tilde{H}^T (\Pi_1^T)^{-1}][\Pi_1^T \tilde{E} - \Pi_1^{-1} \tilde{H}] \geq 0.$$

Lemma 2.3 *Consider the following inequality:*

$$\Gamma(X_1, X_2, \cdots, X_n) - \sum_{i=1}^{n} \digamma_i^T(X_i)\Lambda_i^{-1}(X_i)\digamma_i(X_i) < 0, \tag{2.3}$$

where X_i, $i = 1, \cdots, n$ *are the matrix variables,* $\Lambda_i(X_i) > 0$ *and* $\digamma_i(X_i)$ *are functions of* X_i, $i = 1, \cdots, n$. *Then the inequality in* (2.3) *is feasible in* X_i, $i = 1, \cdots, n$ *if and only if the following inequality is feasible in* X_i, J_i, $i = 1, \cdots, n$:

$$\Gamma(X_1, X_2, \cdots, X_n) + \sum_{i=1}^{n} (J_i^T \Lambda_i(X_i) J_i + J_i^T \digamma_i(X_i) + \digamma_i^T(X_i) J_i) < 0. \tag{2.4}$$

Proof 2 *It can be shown that the feasibility in* X_i, $i = 1, \cdots, n$ *of* (2.3) *is equivalent to the feasibility in* X_i, J_i, $i = 1, \cdots, n$ *of*

$$\begin{aligned} &\Gamma(X_1, X_2, \cdots, X_n) - \sum_{i=1}^{n} \digamma_i^T(X_i)\Lambda_i^{-1}(X_i)\digamma_i(X_i) \\ &+ \sum_{i=1}^{n} ([J_i + \Lambda_i^{-1}(X_i)\digamma_i(X_i)]^T \Lambda_i(X_i)[J_i + \Lambda_i^{-1}(X_i)\digamma_i(X_i)]) < 0, \end{aligned} \tag{2.5}$$

where J_i, $i = 1, \cdots, n$ *are introduced auxiliary variables [86]. Indeed, the inference from* (2.5) *to* (2.3) *is obvious, and the inference from* (2.3) *to* (2.5) *follows by letting* $J_i = -\Lambda_i^{-1}(X_i)\digamma_i(X_i)$. *Then, it is easy to show that* (2.5) *is equivalent to* (2.4). *This completes the proof.*

2.3 Design of Discrete-time SMC

Consider the following linear discrete-time sliding function:

$$\sigma_x(k) = Sx(k), \tag{2.6}$$

where $S \in \mathbb{R}^{m\times n}$ will be designed later such that SB is nonsingular. During the ideal sliding motion the sliding function satisfies:

$$\sigma_x(k+1) = \sigma_x(k) = 0, \ \ \forall k > k_s, \tag{2.7}$$

where $k_s > 0$ denotes the time that sliding motion starts. Thus, one may obtain from (2.1) and (2.6) that

$$\sigma_x(k+1) = S(A+\Delta A_k)x(k) + SB[u(k) + f(k)]. \tag{2.8}$$

Here we will provide the mean value and boundary layer thickness vectors for the exogenous disturbance according to the upper and lower bounds of $f(k)$. In doing so, assume

$$f_i^l \leq f_i(k) \leq f_i^u, \ \ i = 1, \cdots, m, \tag{2.9}$$

where f_i^l and f_i^u denote the lower and upper bound of the i-th entry of $f(k)$. Define

$$f_i^+ = \frac{f_i^u + f_i^l}{2}, \ f_i^- = \frac{f_i^u - f_i^l}{2}, \ \ i = 1, \cdots, m, \tag{2.10}$$

and

$$\mathscr{F}^+ = \mathrm{col}(f_1^+, \cdots, f_m^+), \ \mathscr{F}^- = \mathrm{col}(f_1^-, \cdots, f_m^-), \tag{2.11}$$

where $\mathscr{F}^+$ and $\mathscr{F}^-$ are the mean value and boundary layer thickness vectors of $f(k)$ respectively. Now, the following control law is proposed:

$$u(k) = -(SB)^{-1}SAx(k) - \vartheta(k), \tag{2.12}$$

where $\vartheta(k)$ denotes the approximation of disturbance $f(k)$ which may be used in the controller to compensate the bad effect of disturbance on the ultimate bound on the system state trajectories. $\vartheta(k)$ can also be regarded as the feedforward control, in addition to the linear controller. It is assumed that the component $\vartheta(k)$ is bounded, satisfying

$$\|f(k) - \vartheta(k)\| \leq \tau \left\|\mathscr{F}^-\right\|, \tag{2.13}$$

where τ is a predefined positive scalar depending on the choice of $\vartheta(k)$. More discussions about the component $\vartheta(k)$ and τ are presented later in this chapter.

Remark 2.1 *In this chapter, the control law* (2.12) *uses only the upper and lower bounds on the matched exogenous disturbance. However, it can be seen in the literature ([101], [27], [57]) that the term $S\Delta Ax(k)$ is assumed to be bounded. Broadly speaking, it is not a realistic assumption that we should know a priori some sort of information about the system state bounds. Instead, we eliminate any restrictive bound on the system states and only deal with the system unmatched (mismatched) uncertainties using the robust control strategies.*

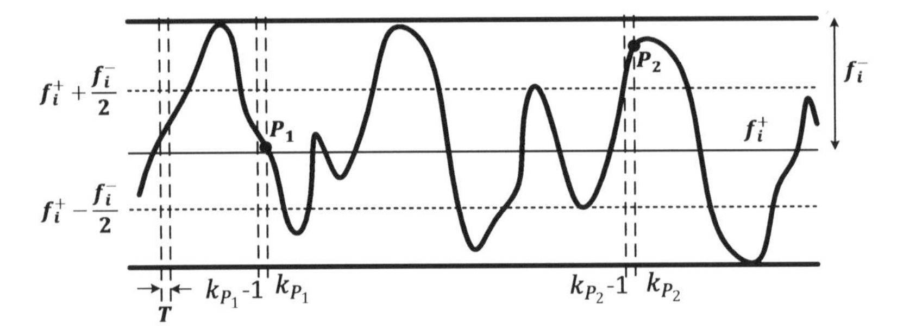

FIGURE 2.1
Signal $f_i(k)$.

2.3.1 Variable structure discontinuous control considerations

As discussed in section 2, in the literature, it is argued that the discontinuous part of the sliding control input can be detrimental to performance [83]. However, this claim is only true for the balanced uncertainties and/or disturbances whose maximum frequency component is close to the sampling rate of the discrete-time system. Specifically, with the smoothness and boundedness conditions of the external disturbance, a number of beneficial choices as discontinuous variable structure components can be utilized in the DSMC in order to improve its performance. To explain, assume that $f_i(k)$ (the ith element in $f(k)$) has the waveform as in Figure 2.1. Now, for instance, to estimate the instantaneous amplitude of disturbance at point P_1, five choices are accessible: 1) *zero*, 2) f_i^+, 3) $f_i^+ + f_i^-$, 4) $f_i^+ + \frac{f_i^-}{2}$, 5) $f_i(k_{P_1} - 1)$. Similarly, for point P_2, one may suggest to use 1) *zero*, 2) f_i^+, 3) $f_i^+ - f_i^-$, 4) $f_i^+ - \frac{f_i^-}{2}$, 5) $f_i(k_{P_2} - 1)$. Here, $f_i(k_{P_j} - 1)$ means the value of f_i at the time instant of $k = k_{P_j} - 1, j = 1,2$. Using the first choice (or indeed the lack of any discontinuous component) in the controller leads to the well-known linear controller. Exploiting the second choice, referred to as the mean value of the exogenous disturbance, in the DSMC has been proposed in [67]. It is presented in [67] that the term f_i^+ can be used in the ith element of the control law to compensate the nonzero mean of unbalanced disturbances. It can easily be realized that in the case of using f_i^+ the maximum estimation error is f_i^-. In the following of this subsection, according to the third and fourth choices, we will discuss two different forms of VSDC for DSMC. The discussion about the last choice, which is referred to as the disturbance observer, will be the subject of the next section. In what follows, we assume that the exogenous disturbance in system (2.1) is smooth and bounded.

Assumption 2.1 *The exogenous disturbance $f(k)$ in (2.1) satisfies the Lipschitz continuity condition and we have,*

$$\|f(k) - f(k-1)\| \leq L_f T_s, \tag{2.14}$$

where $L_f > 0$ denotes Lipschitz constant and T_s is the sampling time.

Here, it will be assumed that L_f has a small value. To this end, the sampling rate of the discrete signal processing system is assumed to be big enough compared to the maximum component frequency of exogenous disturbance $f(k)$. Further in what follows, we assume the known sliding surface matrix S and its design will be derived in Section 2.3.2.

2.3.1.1 Using upper and lower bounds of disturbance in the controller: $\mathscr{C}_1$

Note that

$$f(k-1) = (SB)^{-1}S[x(k) - Ax(k-1) - \Delta A_{k-1}x(k-1) - Bu(k-1)]. \tag{2.15}$$

$f(k-1)$ may be estimated by:

$$\hat{f}(k) = (SB)^{-1}S[x(k) - Ax(k-1) - Bu(k-1)], \tag{2.16}$$

which is equivalent to

$$\hat{f}(k) = (SB)^{-1}S\Delta A_{k-1}x(k-1) + f(k-1).$$

For zero-centered uncertainty ΔA, it is obvious that the term $(SB)^{-1}S\Delta A_{k-1}x(k-1)$ is also zero-centered and has no influence on the mean values of the vector $\hat{f}(k)$. Additionally, in the case that the system state is bounded, the vector $(SB)^{-1}S\Delta A_{k-1}x(k-1)$ remains also bounded. With the proper choice of S and for small uncertainty ΔA, it can be claimed that the magnitude of $(SB)^{-1}S\Delta A_{k-1}x(k-1)$ will be very small compared to $f(k-1)$. Traditionally, *signum* function can be used to determine the position of the instantaneous disturbance relative to the line f_i^+. Hence, one may propose to set $\vartheta(k)$ in (2.12) as:

$$\vartheta_1(k) = \mathscr{F}^+ + \text{diag}(\mathscr{F}^-)\text{sgn}(\hat{f}(k) - \mathscr{F}^+), \tag{2.17}$$

where $\text{diag}(\mathscr{F}^-) := \text{diag}(f_1^-, \cdots, f_m^-)$. Thus, the controller (2.12) can be defined as:

$$u_1(k) = -(SB)^{-1}SAx(k) - \mathscr{F}^+ - \text{diag}(\mathscr{F}^-)\text{sgn}(\hat{f}(k) - \mathscr{F}^+). \tag{2.18}$$

Remark 2.2 *With a quick glimpse into the literature, it can be found that a frequently used candidate for the component $\vartheta(k)$ has the general form of:*

$$\vartheta(k) = \psi + \nu sgn(\sigma_x(k)), \tag{2.19}$$

where ψ and ν are known parameters. For instance, in [101], with ignoring the bounds of $S\Delta A_k x(k)$ (see Remark 2.1), ψ and ν are assumed to be some constants involving the bounds of $SBf(k)$, similar to $\mathscr{F}^+$ and $\mathscr{F}^-$. Regardless of different approaches used to design the parameters of this nonlinear function, it should be emphasized that the term $sgn(\sigma_x(k))$ is not an appropriate function to determine the position of the disturbance relative to its mean value either in the physical meaning or in the theoretical sense. Using the controller containing $\vartheta(k)$ as in (2.19) *will lead state trajectories to chatter around the switching surface with amplitude dependent on the lower bound of the term* (2.19) *and with the frequency equal to the sampling rate; see [67]. Using the controller* (2.18), *while the chattering still happens, in this case, the state trajectories chatter with the frequency equal to the frequency of exogenous disturbance.*

2.3.1.2 Using $f_i^+ \pm \frac{f_i^-}{2}$: $\mathscr{C}_2$

As a new alternative, $f_i^+ \pm \frac{f_i^-}{2}$ can be used as an estimate of P_1 or P_2 in Figure 2.1. The estimation error, in the worst-case scenario, will be $\frac{3}{2}f_i^-$. Hence, one may propose to put the component $\vartheta(k)$ in (2.12) as:

$$\vartheta_2(k) = \mathscr{F}^+ + \frac{1}{2}\text{diag}(\mathscr{F}^-)\text{sgn}(\hat{f}(k) - \mathscr{F}^+). \tag{2.20}$$

Thus, the controller (2.12) is chosen as:

$$u_2(k) = -(SB)^{-1}SAx(k) - \mathscr{F}^+ - \frac{1}{2}\text{diag}(\mathscr{F}^-)\text{sgn}(\hat{f}(k) - \mathscr{F}^+), \tag{2.21}$$

where $\hat{f}(k)$ is defined in (2.16).

2.3.2 Design of a robust sliding surface

The sequel of this section aims to consider the stability of the system (2.1) using the controller (2.12). As a result of applying the controller (2.12) to the system (2.1), it is seen that

$$x(k+1) = (A + \Delta A_k - \hat{A})x(k) + Bf_\vartheta(k), \tag{2.22}$$

where $f_\vartheta(k) \triangleq f(k) - \vartheta(k)$ and $\hat{A} \triangleq B(SB)^{-1}SA$. Furthermore, it can be found that

$$\sigma_x(k+1) = S\Delta A_k x(k) + SBf_\vartheta(k). \tag{2.23}$$

The following lemmas are given to characterize the boundedness of the system state (2.22).

Lemma 2.4 ([77]) *Let $V(\zeta(k))$ be a Lyapunov candidate function. In the case that there exist real scalars $\nu \geq 0$, $\alpha > 0$, $\beta > 0$ and $0 < \rho < 1$ such that*

$$\alpha \|\zeta(k)\|^2 \leq V(\zeta(k)) \leq \beta \|\zeta(k)\|^2,$$

and

$$V(\zeta(k+1)) - V(\zeta(k)) \leq \nu - \rho V(\zeta(k)),$$

then $\zeta(k)$ will satisfy

$$\|\zeta(k)\|^2 \leq \frac{\beta}{\alpha}\|\zeta(0)\|^2(1-\rho)^k + \frac{\nu}{\alpha\rho}.$$

Lemma 2.5 *For any symmetric matrix $P > 0$ and any full column rank matrix B, we have $PB(B^TPB)^{-1}B^TP \leq P$.*

Proof 3 *It can easily be proved by*

$$[I - B(B^TPB)^{-1}B^TP]^TP[I - B(B^TPB)^{-1}B^TP] \geq 0.$$

It should also be noted that with applying DSMC to discrete-time systems involving exogenous disturbances, the closed-loop system should be analyzed in terms of boundedness. Also, the DSMC can only ensure that the state trajectories may be driven into a boundary layer around the ideal sliding surface $\sigma(k) = 0$. This issue is indeed regarded as the quasi sliding mode (QSM) in the literature. On the other hand, due to the presence of mismatched uncertainty in the system dynamics, it is difficult to analyze the reachability of the QSM by means of a separate sufficient condition. Alternatively, the following theorem considers a method to analyze simultaneously the reachabiltiy of QSM and the stability of the system states by means of a discrete-time Lyapunov stability method.

Theorem 2.1 *In the absence of disturbance $f(k)$, the linear part of the control law* (2.12) *can drive the system state onto the ideal sliding surface* (2.6)*, and the system state is stabilized, if there exist a symmetric matrix $\bar{P} > 0$, matrices X and Y, and scalars $\varepsilon > 0$ and $\bar{\eta} > 0$ satisfying the following LMI:*

$$\begin{bmatrix} -\bar{P}+Y^TB^T+BY & \star & \star & \star & \star & \star \\ 0 & -\bar{P}+2\varepsilon MM^T & \star & \star & \star & \star \\ A\bar{P}+BX & \sqrt{2}\varepsilon MM^T & -\bar{P}+\varepsilon MM^T & \star & \star & \star \\ BY & 0 & 0 & -\bar{P} & \star & \star \\ \bar{P} & 0 & 0 & 0 & -\bar{\eta}I & \star \\ N\bar{P} & 0 & 0 & 0 & 0 & -\varepsilon I \end{bmatrix} < 0, \tag{2.24}$$

where M and N are known matrices in (2.2)*. Here $S = B^T\bar{P}^{-1}$ and $\{\star\}$ denotes the symmetric elements in a symmetric matrix.*

Proof 4 *Define*

$$V(\zeta(k)) = x^T(k)Px(k) + \sigma_x^T(k)(SB)^{-1}\sigma_x(k), \tag{2.25}$$

where $\zeta(k) = \begin{bmatrix} x^T(k) & \sigma_x^T(k) \end{bmatrix}^T$, $P > 0$ is a symmetric matrix and $S = B^TP$. Thus, we can write

$$\begin{aligned} \Delta V(\zeta(k)) =& V(\zeta(k+1)) - V(\zeta(k)) \\ =& x^T(k+1)Px(k+1) + \sigma_x^T(k+1)(SB)^{-1}\sigma_x(k+1) \\ & - x^T(k)Px(k) - \sigma_x^T(k)(SB)^{-1}\sigma_x(k). \end{aligned} \tag{2.26}$$

Now, it can be shown that

$$\Delta V(\zeta(k)) = \begin{bmatrix} x(k) \\ f_\vartheta(k) \end{bmatrix}^T \begin{bmatrix} \Omega_{11} & \Omega_{12} \\ \Omega_{12}^T & \Omega_{22} \end{bmatrix} \begin{bmatrix} x(k) \\ f_\vartheta(k) \end{bmatrix}, \tag{2.27}$$

where

$$\begin{aligned} \Omega_{11} :=& (A+\Delta A_k)^T P(A+\Delta A_k) - (A+\Delta A_k)^T PB(B^TPB)^{-1}B^TP(A+\Delta A_k) \\ & - P - PB(B^TPB)^{-1}B^TP + 2\Delta A_k^T PB(B^TPB)^{-1}B^TP\Delta A_k, \\ \Omega_{12} :=& 2\Delta A_k^T S^T, \\ \Omega_{22} :=& 2SB, \end{aligned}$$

In the absence of the disturbance $f(k)$, that is $f(k)=0$, thus $\vartheta(k)=0$ leading to $f_\vartheta(k)=0$. Then the system is stabilized if

$$\Omega_{11} < -\eta I, \tag{2.28}$$

where $\eta > 0$ is a scalar variable. Now, we consider the feasibility of (2.28). *To obtain* (2.28), *by utilizing Lemma 2.5 and the Schur complement, it suffices to have*

$$\begin{bmatrix} \hat{\Omega}_{11} & \sqrt{2}\Delta A_k^T \\ \sqrt{2}\Delta A_k & -\bar{P} \end{bmatrix} < 0, \tag{2.29}$$

where $\bar{P}=P^{-1}$, and

$$\begin{aligned}\hat{\Omega}_{11} =&(A+\Delta A_k)^T P(A+\Delta A_k)-(A+\Delta A_k)^T PB(B^T PB)^{-1}B^T P(A+\Delta A_k)\\ &-P-PB(B^T PB)^{-1}B^T P+\eta I.\end{aligned}$$

According to Lemma 2.3, the feasibility of the inequality in (2.29) *is equivalent to that of*

$$\begin{bmatrix} \tilde{\Omega}_{11} & \sqrt{2}\Delta A_k^T \\ \sqrt{2}\Delta A_k & -\bar{P} \end{bmatrix} < 0, \tag{2.30}$$

where

$$\begin{aligned}\tilde{\Omega}_{11} =&(A+\Delta A_k+BF)^T P(A+\Delta A_k+BF)-P+\eta I\\ &+L^T(B^T PB)L+L^T B^T P+PBL.\end{aligned}$$

Here, F and L are two auxiliary variables [86]. Then, by left and right matrix multiplication on both sides of the inequality in (2.30) *with diag$(\bar{P}, I)$, we have*

$$\begin{bmatrix} \bar{P}\tilde{\Omega}_{11}\bar{P} & \sqrt{2}\bar{P}\Delta A_k^T \\ \sqrt{2}\Delta A_k\bar{P} & -\bar{P} \end{bmatrix} < 0. \tag{2.31}$$

Using the Schur complement and Lemma 2.1, it can be demonstrated that the inequality in (2.31) *can be implied by the LMI in* (2.24), *where $X=F\bar{P}$, $Y=L\bar{P}$ and $\bar{\eta}=\eta^{-1}$.*

Remark 2.3 *It is worth mentioning that as $\zeta=[x^T\ \sigma^T]^T=\begin{bmatrix} I \\ B^T P\end{bmatrix}x$ and $\begin{bmatrix} I \\ B^T P\end{bmatrix}$ is a full rank matrix, $\zeta=0$ if and only if $x=0$. In addition, a key feature in our method to prove the above theorem (and Theorem 2.3 later in the chapter), and further design the sliding function matrix S, is to neglect the bounded inputs (e.g., the nonlinear control and exogenous disturbance), and directly prove the stability of the unforced linear system. More precisely, from* (2.27) *(with $f_\vartheta(k)=0$) and* (2.28) *we may write*

$$\begin{aligned}\Delta V(\zeta) :=& x^T(k)\Omega_{11}x(k)\\ \le& -\eta x^T(k)x(k)\\ <& -\frac{\eta}{\lambda_{\max}(P+PB(B^T PB)^{-1}B^T P)}x^T(k)\{P+PB(B^T PB)^{-1}B^T P\}x(k)\\ \triangleq& -\rho V(\zeta),\end{aligned}$$

which ensures the asymptotic stability of the closed-loop system and thus $\zeta \to 0$, $\sigma \to 0$ *and* $x \to 0$.

Remark 2.4 *The proof of this theorem provides a less conservative sufficient condition for the design of a robust sliding matrix for the system in* (2.1) *involving mismatched uncertainties. Further based on this proof, the second objective of this chapter, when the disturbance estimator is utilized in the controller directly, will be derived in the proof of Theorem 2.3.*

2.3.3 Characterizing system state boundedness

While Theorem 2.1 presents a method to design the DSMC in order to stabilize the system in (2.1), it does not present a bound on the system states. The following theorem characterizes the boundedness of the obtained closed-loop system state and corresponding sliding function.

Theorem 2.2 *In the presence of disturbance* $f(k)$, *if the LMI in* (2.24) *is feasible, for the obtained* $P = \bar{P}^{-1}$ *and* $\eta = \bar{\eta}^{-1}$, *the controller* (2.12) *satisfying* (2.13) *will lead to a bound on the augmented system state* $\zeta(k) = [x^T(k), \sigma_x^T(k)]^T$ *as follows:*

$$\begin{aligned} &\forall \varsigma > 0,\ \exists k^\star > 0,\ s.t.\ \forall k > k^\star, \\ &\|\zeta(k)\|^2 \leq \frac{\lambda_{\max}(\mathbf{M})}{\hat{\eta}\lambda_{\min}(diag(P,(B^TPB)^{-1}))}\gamma + \varsigma, \end{aligned} \tag{2.32}$$

where $\mathbf{M} = PB(B^TPB)^{-1}B^TP + P$, *and* $\gamma = \tau^2 \left\|\mho + 2B^TPB\right\| \|\mathscr{F}^-\|^2$; *here the scalar variable* $\hat{\eta} > 0$ *and matrix variable* $\mho > 0$ *are obtained from solving the following LMI:*

$$\begin{bmatrix} (\hat{\eta}-\eta)I + 4\bar{\varepsilon}N^TN & \star & \star \\ 0 & -\mho & \star \\ 0 & M^TPB & -\bar{\varepsilon}I \end{bmatrix} < 0, \tag{2.33}$$

where M and N are known matrices in (2.2), *and further,* $\bar{\varepsilon} > 0$ *is a scalar variable.*

Proof 5 *According to Lemma 2.2 it can be written that*

$$2x^T(k)\Omega_{12}f_\vartheta(k) \leq x^T(k)\Omega_{12}\mho^{-1}\Omega_{12}^Tx(k) + f_\vartheta^T(k)\mho f_\vartheta(k), \tag{2.34}$$

where $\mho > 0$. *It follows from* (2.27), (2.28) *and* (2.34) *that*

$$\begin{aligned} \Delta V(\zeta(k)) \leq & -\tilde{x}^T(k)\left\{\eta I - \Omega_{12}\mho^{-1}\Omega_{12}^T\right\}\tilde{x}(k) \\ & + f_\vartheta^T(k)[\mho + \Omega_{22}]f_\vartheta(k). \end{aligned} \tag{2.35}$$

If we choose $\mho > 0$ *such that*

$$\hat{\eta}I < \eta I - \Omega_{12}\mho^{-1}\Omega_{12}^T, \tag{2.36}$$

where $0 < \hat{\eta} < \eta$, *which is always possible if* $\eta > 0$ *exists, then it follows from Equation* (2.35) *that*

$$\Delta V(\zeta(k)) \leq -\hat{\eta} x^T(k)x(k) + f_\vartheta(k)^T[\mho + \Omega_{22}]f_\vartheta(k). \tag{2.37}$$

Note also that

$$\begin{aligned} V(\zeta(k)) =& x^T(k)[P + PB(B^T PB)^{-1} B^T P]x(k) \\ \triangleq& x^T(k)\mathbf{M}x(k), \end{aligned} \tag{2.38}$$

hence,

$$\lambda_{\min}(\mathbf{M}) \|\tilde{x}(k)\|^2 \leq V(\zeta(k)) \leq \lambda_{\max}(\mathbf{M}) \|\tilde{x}(k)\|^2. \tag{2.39}$$

Furthermore, it can be shown that

$$\begin{aligned} \lambda_{\min}(diag(P,(B^T PB)^{-1})) \|\zeta(k)\|^2 \leq V(\zeta(k)) \\ \leq \lambda_{\max}(diag(P,(B^T PB)^{-1})) \|\zeta(k)\|^2, \end{aligned} \tag{2.40}$$

Therefore, from (2.37) *and* (2.39) *one can derive that*

$$\Delta V(\zeta(k)) \leq -\frac{\hat{\eta}}{\lambda_{\max}(\mathbf{M})} V(\zeta(k)) + \gamma, \tag{2.41}$$

where, due to the continuity assumption mentioned in (2.14),

$$\gamma = \tau^2 \|\mho + 2B^T PB\| \|\mathscr{F}^-\|^2.$$

Note that from (2.27) *it can simply be written that*

$$x^T(k)\Omega_{11}x(k) = V(\zeta(k+1))\big|_{f_\vartheta(k)=0} - V(\zeta(k)) < -\eta x^T(k)x(k). \tag{2.42}$$

It is known that $V(\zeta(k+1))\big|_{f_\vartheta(k)=0} \geq 0$, *and thus, from* (2.42) *and* (2.39), *it can be claimed that* $\lambda_{\max}(\mathbf{M}) > \eta$. *Therefore*

$$\frac{\hat{\eta}}{\lambda_{\max}(\mathbf{M})} < 1.$$

Thus, from Lemma 2.4, (2.40) *and* (2.41), *the bound in* (2.32) *can be obtained.*

Now let us consider how to solve the inequality (2.36). *By the aid of Lemma 2.1 and the Schur complement, it can be shown that for the given* $P > 0$ *and* $\eta > 0$, *this inequality can be implied by the LMI in* (2.33).

To be more specific, if one utilizes the controller in (2.18) ($\mathscr{C}_1$), $\tau_1 = 2$ in (2.13) and $\gamma_1 = 4\|\mho + 2SB\| \|\mathscr{F}^-\|^2$ in (2.32). Note that this bound results from the worst case scenario. However, if the signum function $\text{sgn}(\hat{f}(k) - \mathscr{F}^+)$ can predict perfectly the location of $f(k)$, which is assumed to be the most cases for slow disturbances, this bound can be reduced to $\tau_1^\star = 1$ and $\gamma_1^\star = \|\mho + 2SB\| \|\mathscr{F}^-\|^2$.
On the other hand, utilizing the controller in (2.21)($\mathscr{C}_2$), we have $\tau_2 = 1.5$ and $\gamma_2 = 2.25\|\mho + 2SB\| \|\mathscr{F}^-\|^2$. It should be noted that this bound is also the worst case scenario bound. Since it is assumed that disturbance in the system (2.1) is slow, this bound, with perfect position estimation, can be reduced to $\tau_2^\star - 0.5$ and $\gamma_2^\star = 0.25\|\mho + 2SB\| \|\mathscr{F}^-\|^2$.

2.4 Exploiting Disturbance Estimate in the Control Law: $\mathscr{C}_3$

According to the paper [133], for smooth disturbances, $f(k-1)$ is a good approximation to $f(k)$ so as to reduce the ultimate bound on the system state. But, unlike [133] in which the system is not uncertain and just involves exogenous disturbance, in this chapter we consider a discrete-time system involving uncertainty and exogenous disturbance. Due to the presence of system uncertainty, as seen in (2.15), we do not have direct access to $f(k-1)$, thus $\hat{f}(k)$ in (2.16) is used here instead. Furthermore, using the term $\hat{f}(k)$ in the controller directly, rather than using the ones proposed previously seems to have much better performance. Now, by substituting

$$\vartheta_3(k) = \hat{f}(k) \tag{2.43}$$

in (2.12), the following controller is achieved

$$u_3(k) = -(SB)^{-1}SAx(k) - \hat{f}(k). \tag{2.44}$$

A similar idea can also be found in e.g. [133]. Note that, referring to (2.16), $\vartheta(k)$ in (2.43) includes system uncertainty, and thus the condition in (2.13) does not apply. Therefore, the stability of the closed-loop system should be analyzed again. In the following, we consider the stability of the system (2.1) using the controller (2.44).

By applying the controller (2.44) to the system (2.1), we have

$$x(k+1) = (A + \Delta A_k - \hat{A})x(k) - B(SB)^{-1}S\Delta A_{k-1}x(k-1) + Bf_d(k), \tag{2.45}$$

where $f_d(k) \triangleq f(k) - f(k-1)$ and $\hat{A} \triangleq B(SB)^{-1}SA$. As seen, the closed-loop system (2.45) involves time-delay. Furthermore, it can be found that

$$\sigma_x(k+1) = S\Delta A_k x(k) - S\Delta A_{k-1}x(k-1) + SBf_d(k). \tag{2.46}$$

Theorem 2.3 *In the absence of disturbance $f(k)$, the control law* (2.44), (2.16) *can drive the system state onto the ideal sliding surface* (2.6) *and the system state is stabilized, if there exist symmetric matrices $\bar{P} > 0$ and $\bar{Q} > 0$, matrices X and Y and also scalars $\varepsilon_1 > 0$, $\varepsilon_2 > 0$ and $\eta_1 > 0$ satisfying the following LMI:*

$$\begin{bmatrix} \hat{\mathscr{M}}_{11} & \star & \star & \star & \star & \star & \star & \star \\ 0 & -\bar{Q} & \star & \star & \star & \star & \star & \star \\ 0 & 0 & \hat{\mathscr{M}}_{22} & \star & \star & \star & \star & \star \\ A\bar{P}+BX & 0 & \sqrt{2}\varepsilon_1 MM^T & -\bar{P}+\varepsilon_1 MM^T & \star & \star & \star & \star \\ BY & 0 & 0 & 0 & -\bar{P} & \star & \star & \star \\ \bar{P} & 0 & 0 & 0 & 0 & -\bar{\eta}_1 I & \star & \star \\ N\bar{P} & 0 & 0 & 0 & 0 & 0 & -\varepsilon_1 I & \star \\ 0 & N\bar{P} & 0 & 0 & 0 & 0 & 0 & -\varepsilon_2 I \end{bmatrix} < 0 \tag{2.47}$$

where $\hat{\mathscr{M}}_{11} = -\bar{P} + \bar{Q} + Y^T B^T + BY$, $\hat{\mathscr{M}}_{22} = -\bar{P} + 2(\varepsilon_1 + \varepsilon_2)MM^T$. Here M and N are known matrices in (2.2), *and $S = B^T\bar{P}^{-1}$.*

Proof 6 *Define*

$$V(\zeta(k)) = x^T(k)Px(k) + x^T(k-1)Qx(k-1) + \sigma_x^T(k)(SB)^{-1}\sigma_x(k),$$

where $\zeta(k) = \begin{bmatrix} x^T(k) & x^T(k-1) & \sigma_x^T(k) \end{bmatrix}^T$, $P > 0$ *and* $Q > 0$ *are symmetric matrices and* $S = B^T P$. *Thus, we can write*

$$\begin{aligned}\Delta V(\zeta(k)) =& V(\zeta(k+1)) - V(\zeta(k)) \\ =& x^T(k+1)Px(k+1) + x^T(k)Qx(k) + \sigma_x^T(k+1)(SB)^{-1}\sigma_x(k+1) \\ & - x^T(k)Px(k) - x^T(k-1)Qx(k-1) - \sigma_x^T(k)(SB)^{-1}\sigma_x(k). \qquad (2.48)\end{aligned}$$

Now, it can be shown that

$$\Delta V(\zeta(k)) = \begin{bmatrix} x(k) \\ x(k-1) \\ f_d(k) \end{bmatrix}^T \begin{bmatrix} \Sigma_{11} & \Sigma_{12} & \Sigma_{13} \\ \Sigma_{12}^T & \Sigma_{22} & \Sigma_{23} \\ \Sigma_{13}^T & \Sigma_{23}^T & \Sigma_{33} \end{bmatrix} \begin{bmatrix} x(k) \\ x(k-1) \\ f_d(k) \end{bmatrix}, \qquad (2.49)$$

where

$$\begin{aligned}\Sigma_{11} :=& (A+\Delta A_k)^T P(A+\Delta A_k) - (A+\Delta A_k)^T PB(B^T PB)^{-1} B^T P(A+\Delta A_k) \\ & - P + Q - PB(B^T PB)^{-1} B^T P + 2\Delta A_k^T PB(B^T PB)^{-1} B^T P \Delta A_k, \\ \Sigma_{12} :=& -2\Delta A_k^T S^T (SB)^{-1} S \Delta A_{k-1}, \\ \Sigma_{22} :=& 2\Delta A_{k-1}^T PB(B^T PB)^{-1} B^T P \Delta A_{k-1} - Q,\end{aligned}$$

and $\Sigma_{13} = 2\Delta A_k^T S^T$, $\Sigma_{23} = -2\Delta A_{k-1}^T S^T$ *and* $\Sigma_{33} = 2SB$. *In the absence of the disturbance* $f(k)$, $f_d(k) = 0$. *Then the system is stabilized if*

$$\Upsilon := \begin{bmatrix} \Sigma_{11} & \Sigma_{12} \\ \Sigma_{12}^T & \Sigma_{22} \end{bmatrix} < -\eta_1 I, \qquad (2.50)$$

where $\eta_1 > 0$ *is a scalar variable. Following a similar approach given in the proof of Theorem 2.1, and by using the Schur complement, Corollary 2.1, Lemma 2.3 and Lemma 2.5, it can be demonstrated that the inequality in* (2.50) *can be implied by the LMI in* (2.47), *where* $\bar{Q} = \bar{P}Q\bar{P}$, $X = F\bar{P}$, $Y = L\bar{P}$ *(F and L are two auxiliary variables), and* $\bar{\eta}_1 = \eta_1^{-1}$.

It should be pointed out that the above theorem provides a method to design the disturbance observer based DSMC in order to stabilize the system in (2.1). However, it does not give a bound on the system states. The following theorem characterizes the boundedness of the obtained closed-loop system state and associated sliding function.

Theorem 2.4 *In the presence of disturbance* $f(k)$ *satisfying* (2.14), *if the LMI in* (2.47) *is feasible, for the obtained* $P = \bar{P}^{-1}$, $Q = P\bar{Q}P$, $\eta_1 = \bar{\eta}_1^{-1}$, *the control* *law* (2.44), (2.16) *will lead to a bound on the augmented system state* $\zeta(k) = [x^T(k), x^T(k-1), \sigma_x^T(k)]^T$ *as follows:*

$$\forall \upsilon > 0,\ \exists k^\star > 0,\ s.t.\ \forall k > k^\star,$$
$$\|\zeta(k)\|^2 \le \frac{\lambda_{\max}(diag(\mathbf{M}, Q))}{\hat{\eta}_1 \lambda_{\min}(diag(P, Q, (B^T PB)^{-1}))}\hat{\gamma} + \upsilon, \tag{2.51}$$

where $\mathbf{M} = PB(B^T PB)^{-1}B^T P + P$, *and* $\hat{\gamma} = \left\|\hat{\mho} + 2B^T PB\right\| L_f^2 T_s^2$; *here the scalar variable* $\hat{\eta}_1 > 0$ *and matrix variable* $\hat{\mho} > 0$ *are obtained from solving the following LMI:*

$$\begin{bmatrix} (\hat{\eta}_1 - \eta_1)I + 4\hat{\varepsilon}_1 N^T N & \star & \star & \star & \star \\ 0 & (\hat{\eta}_1 - \eta_1)I + 4\hat{\varepsilon}_2 N^T N & \star & \star & \star \\ 0 & 0 & -\hat{\mho} & \star & \star \\ 0 & 0 & M^T PB & -\hat{\varepsilon}_1 I & \star \\ 0 & 0 & M^T PB & 0 & -\hat{\varepsilon}_2 I \end{bmatrix} < 0, \tag{2.52}$$

where M *and* N *are known matrices in* (2.2), $\hat{\varepsilon}_1 > 0$ *and* $\hat{\varepsilon}_2 > 0$ *are scalar variables.*

Proof 7 *The proof of this theorem is an application of the proof of Theorem 2.2 and thus is omitted here for the purpose of brevity.*

Remark 2.5 *As seen, applying the controller* $\mathscr{C}_3$ *to the system* (2.1) *results in* $\hat{\gamma}_3 = \left\|\hat{\mho} + 2B^T PB\right\| L_f^2 T_s^2$ *in* (2.51). *Obviously, due to the much smaller* $L_f^2 T_s^2$ *in* $\hat{\gamma}_3$, *which is of* $O(T_s^2)$, *the thickness of the boundary layer is reduced, compared to its previous counterparts which are of* $O(T_s)$, *for the smooth disturbance* $f(k)$ *satisfying* (2.14).

2.5 Simulation Results

Once again let us consider the double integrator system (1.1), (1.3) in Chapter 1. Note that the system in (1.1), (1.3) can for example be used to describe the behavior of a single-dimensional motion of a unit mass (see Fig. 2.2). In order to achieve a discrete-time system, we descretize the system (1.1), (1.3) using a sampling time of 0.1 s, and the result is

$$A = \begin{bmatrix} 1.0000 & 0.1000 \\ 0 & 1.0000 \end{bmatrix}, B = \begin{bmatrix} 0.0100 \\ 0.1000 \end{bmatrix} \tag{2.53}$$

We also consider the following uncertainty parameters and disturbance in the system

$$M = \begin{bmatrix} 0.0500 & 0.1500 \end{bmatrix}^T,\ N = \begin{bmatrix} -0.0500 & 0.1000 \end{bmatrix},$$
$$R(k) = 0.3\sin(k).$$

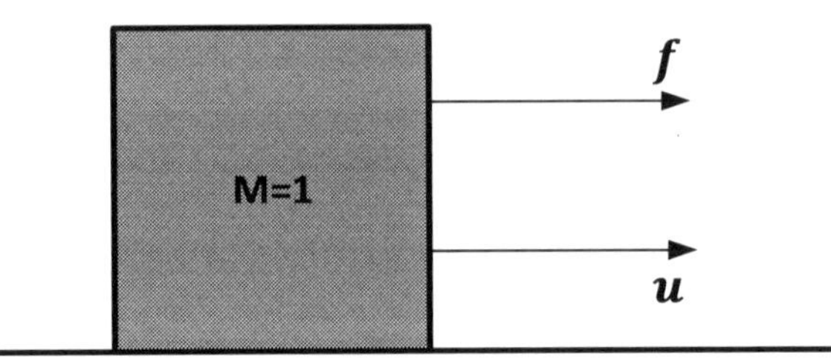

FIGURE 2.2
Plant: Single dimensional motion of a unit mass.

Suppose

$$f(k) = 0.5\left(2 - \sin\left(\frac{k}{5}\right)\right).$$

Solving the LMI (2.24) gives the following results:

$$\bar{P} = \begin{bmatrix} 4.5023 & -1.3481 \\ -1.3481 & 28.4716 \end{bmatrix}, \; \bar{\eta} = 58.2374, \; \varepsilon = 27.6184,$$
$$S = \begin{bmatrix} 0.0033 & 0.0037 \end{bmatrix}.$$

Hence, using $P = \bar{P}^{-1}$, $\mathscr{F}^+ = 1$ and $\mathscr{F}^- = 0.5$, the control laws $\mathscr{C}_1$ and $\mathscr{C}_2$ given in (2.18) and (2.21), respectively, are obtained. The results by applying these controllers, in addition to the linear controller and the DSMC utilizing only the mean value of the disturbance, to the system (2.1) are shown in Figs. 2.3-2.6. Here, the initial state is assumed to be $x(0) = \begin{bmatrix} 2 & -1 \end{bmatrix}^T$. It can be seen that the system state is bounded and also during the sliding motion the state trajectories are within a boundary layer around the sliding surface $\sigma_x(k) = 0$.

As seen, for the slow disturbance $f(k)$, in terms of ultimate bound on the system state and also thickness of the boundary layer around the ideal sliding surface, among these four controllers, the controller $\mathscr{C}_2$ has the best performance.

As mentioned in Remark 2.2, in [101] the following control law is proposed:

$$u(k) = -(SB)^{-1}SAx(k) - \mathscr{F}^+ - \text{diag}(\mathscr{F}^-)\text{sgn}(\sigma_x(k)). \tag{2.54}$$

Figure 2.7 shows the results by applying this controller to the system (2.1). Note that as the LMI condition in [101] is not feasible, the controller (2.54) is constructed by the choice of S achieved through solving the LMI in (2.24). This indeed shows the superiority of our approaches compared to the existing literature. As it is mentioned in [67], this controller leads state trajectories to chatter around the sliding surface with amplitude dependent on the lower bound of the component in (2.19) and with the frequency equal to the sampling rate. It is however discussed in Remark 2.2 that while by using the controllers $\mathscr{C}_1$ and $\mathscr{C}_2$, the chattering issue still exists, in this case,

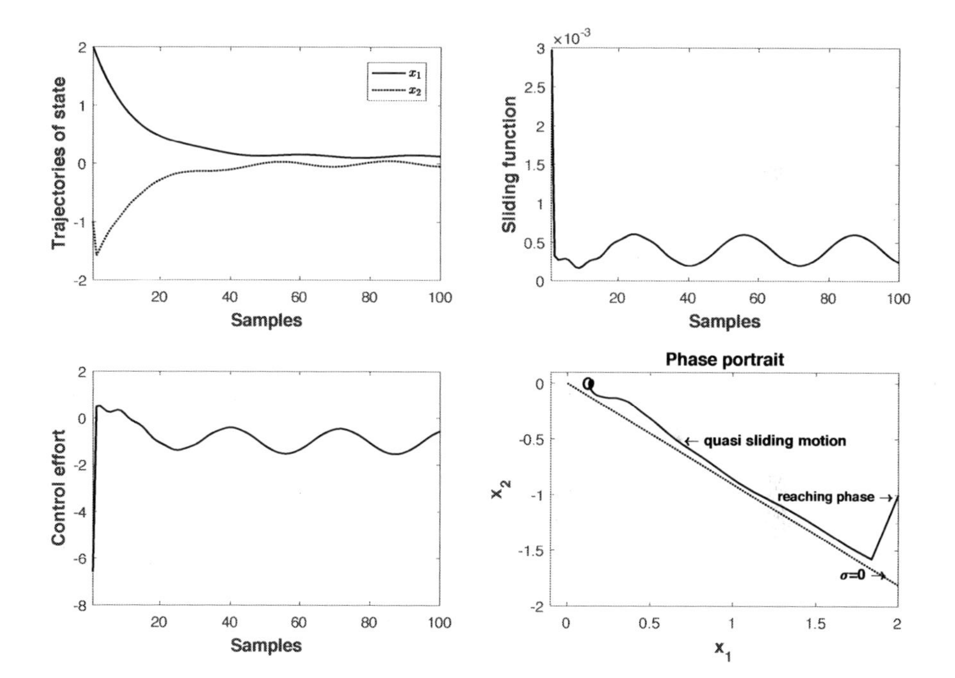

FIGURE 2.3
Results of linear controller.

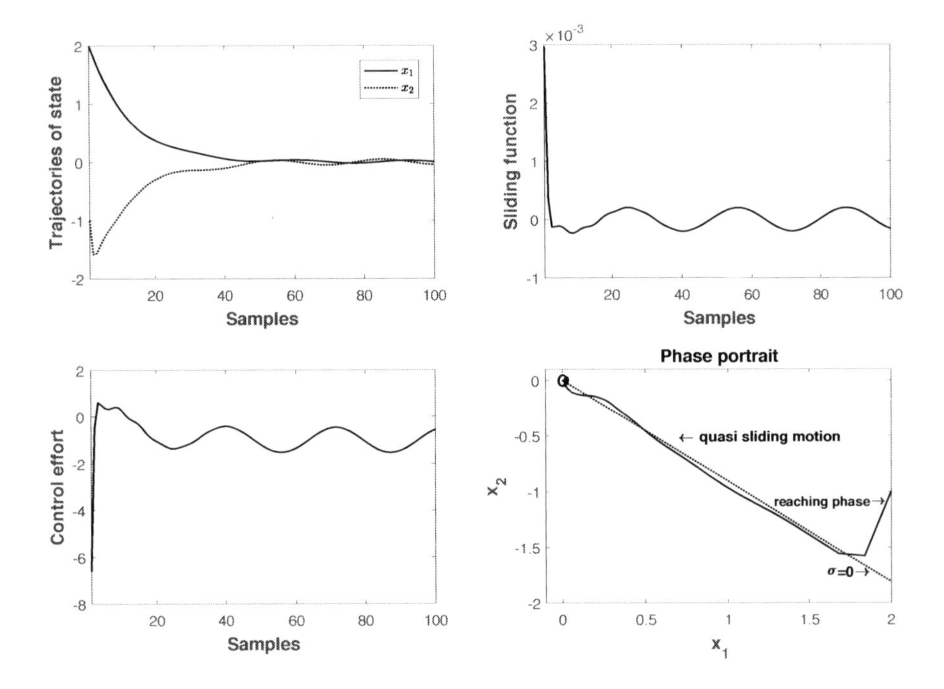

FIGURE 2.4
Results of linear controller and using mean value of disturbance in DSMC.

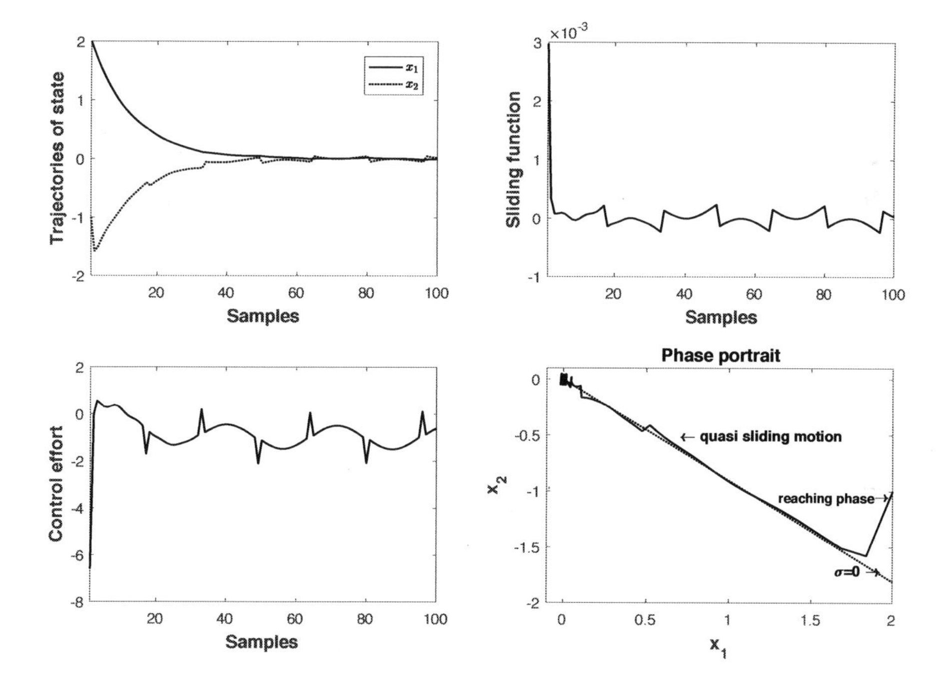

FIGURE 2.5
Results of controller $\mathscr{C}_1$.

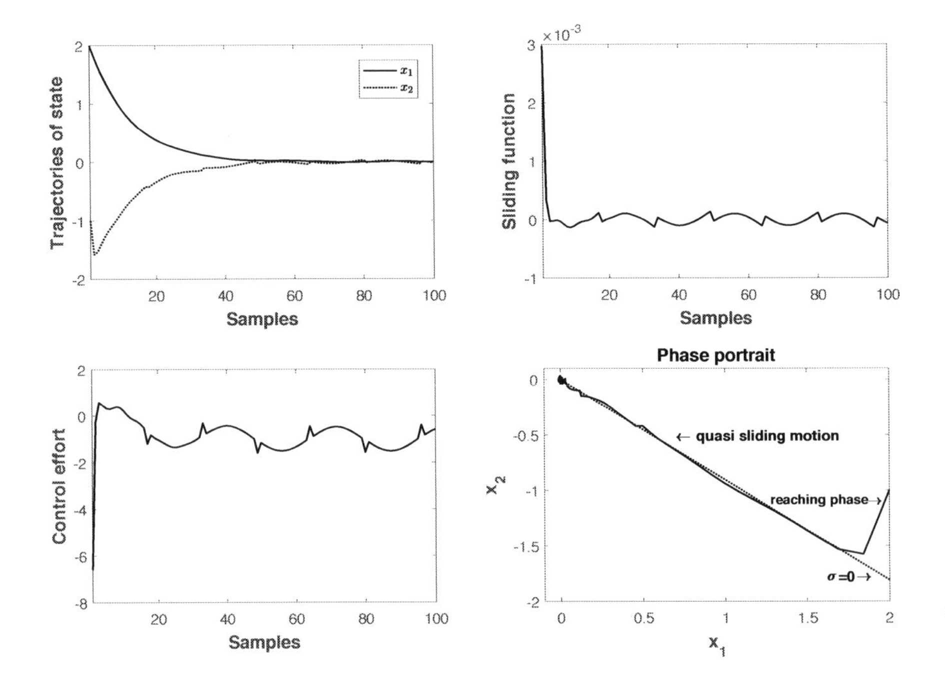

FIGURE 2.6
Results of controller $\mathscr{C}_2$.

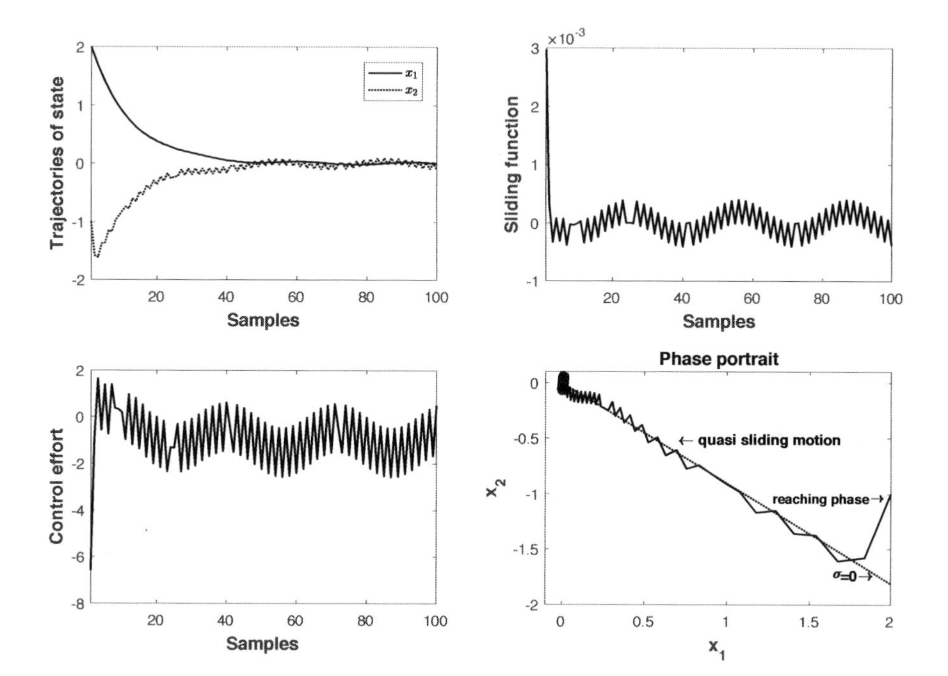

FIGURE 2.7
Results of the controller in [101].

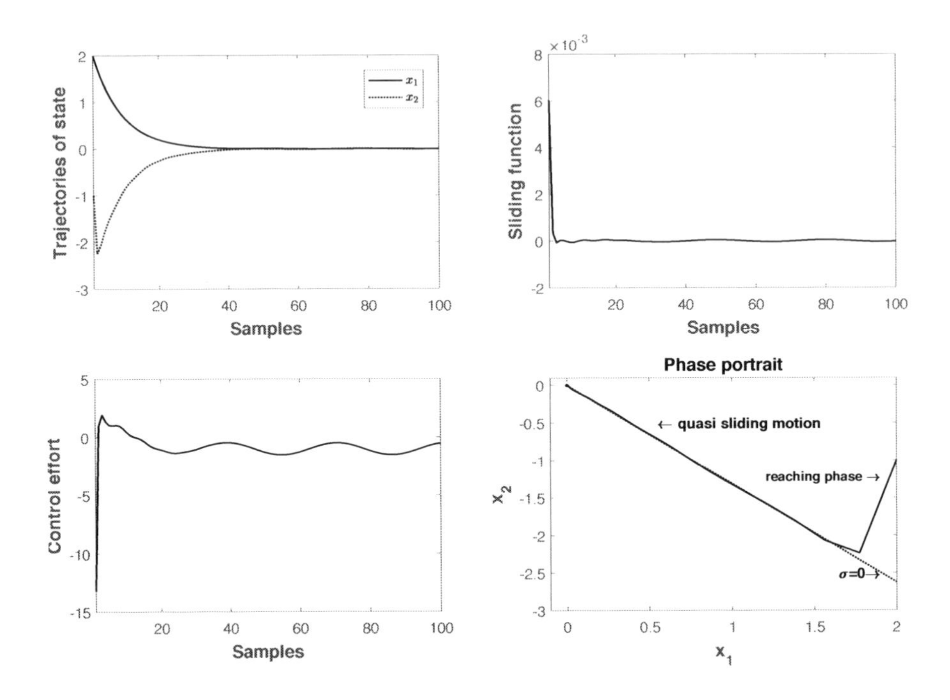

FIGURE 2.8
Results of controller $\mathscr{C}_3$.

the state trajectories chatter with the frequency equal to the frequency of exogenous disturbance.

Then we solve the LMI in (2.47) to design the sliding matrix S and the following results are obtained:

$$\bar{P} = \begin{bmatrix} 2.0996 & -0.0645 \\ -0.0645 & 26.9917 \end{bmatrix}, \ \bar{Q} = \begin{bmatrix} 0.2672 & 1.3040 \\ 1.3040 & 26.0741 \end{bmatrix},$$

$$\bar{\eta}_1 = 61.2774, \ \varepsilon_1 = 26.2248, \ \varepsilon_2 = 34.9235,$$
$$S = \begin{bmatrix} 0.0049 & 0.0037 \end{bmatrix}.$$

The results of applying the control law $\mathscr{C}_3$ in (2.44) to the system (2.1) are illustrated in Fig. 2.8. As it is evident from these results, this controller has the best performance compared to the previous controllers, in terms of ultimate bound on system state as well as boundary layers thickness, for the systems involving smooth disturbances.

2.6 Conclusions

In this chapter, a new LMI based robust DSMC for the systems involving unmatched uncertainty and matched disturbance has been developed. The proposed LMI method is applicable to general systems including unstable systems. Furthermore, some notes on the use of the discontinuous term in the discrete-time sliding mode controller have been given and two new switching functions have been developed. Inspired by the idea of a disturbance observer, a new controller for the underlying uncertain systems has been proposed in this chapter. The controller with disturbance estimator outperforms the other kinds of DSMCs, including linear controller and DSMCs using discontinuous components, while the underlying systems involve slow exogenous disturbances. Nevertheless, the downside is that, since the order of the closed-loop system increases, the scheme using DSMC with disturbance estimator is more computationally intensive.

3

LMI-based output feedback DSMC

CONTENTS

3.1 Introduction 51
3.2 Problem Formulation 52
3.3 Observer-Based Output Feedback DSMC 54
3.3.1 Stability analysis 55
3.3.2 Characterizing the system state boundedness 59
3.3.3 Discussions 61
3.4 Simulation Results 62
3.5 Conclusions 65

Abstract– This chapter is devoted to the problem of designing a robust dynamic output feedback discrete-time sliding mode control (ODSMC) for uncertain systems. The basic idea behind the proposed scheme comes from the fact that ODSMC, unlike its continuous-time counterpart, does not require exploiting a discontinuous term including the sliding function, and hence, it is not a vital requirement that the sliding function is expressed in terms of the outputs only. Less conservative LMI techniques are utilized leading to an LMI condition whose feasible region is greatly wider compared to the existing literature. Also with the assumption of dealing with systems involving slow exogenous disturbances, this scheme can reduce the thickness of the boundary layer around the sliding surface and the ultimate bound on the system state by exploiting a disturbance estimator in the controller rather than discontinuous component. Furthermore, the boundedness of the obtained closed-loop system is analyzed and the bound on the underlying system state is derived.

3.1 Introduction

Sliding mode control has been designed for the cases that all the system states are available. This is not very realistic for most practical problems [83]. Hence, this fact has motivated the researchers to design controllers which exploit only the available information. The literature which has explored output feedback discrete-time sliding mode control (ODSMC) includes both the dynamic and static output feedback controllers [83, 85, 134, 97, 57]. Reference [42] proposes an observer-based sliding mode controller for continuous-time MIMO systems. Different frameworks and discussions for the design of a static output feedback sliding mode controller are given in [57, 43, 62]. Moreover, in order to design direct torsion control of flexible shaft, [81] develops an observer-based discrete-time sliding mode control (DSMC) scheme.

According to the results presented in [67, 127], the use of a switching component in the control law may not necessarily improve the performance. Thanks to the fact that the sliding function is not required to be exploited in the ODSMC, this chapter considers a sliding surface in the state space rather than state estimate space or state estimation error space [103] and [147]. This fact makes it possible to develop a considerably less conservative LMI condition, which is prepared in this chapter to select the switching function matrix. Hence, the feasibility region of the LMI condition obtained in the proposed *scheme*, or equivalently its applicability region, is interestingly improved compared to that presented in the literature; *Cf.* [103] and [147].

Specifically, the proposed *scheme*, with the smoothness and boundedness conditions of the external disturbance, exploits a disturbance estimator in the controller rather than VSDC. Notice that disturbance observer-based control strategies have been exploited in different fields in the literature and have been successfully implemented for different aims; see e.g. [87]. This idea has been extended to the DSMC in [133] in order to reduce the boundary layer thickness. However, the disturbance estimator in Chapter 2 has been designed for the cases that the system states are entirely available and the system does not involve unmatched uncertainties. A framework by exploiting output information only for discrete-time MIMO systems with unmatched disturbances and without uncertainties has been proposed in [32]. Indeed, the idea is to use an integral term of the estimation output error, in addition to the well-known Luenberger observer which observes the system state with a proportional loop, to make more degrees of freedom. This matter is referred to as *proportional integral observer* (PIO) in the literature [32]. Nevertheless, the underlying system in [32], unlike the system considered in this chapter, does not involve unmatched uncertainties. The proposed *scheme* here extends the problem of utilizing disturbance observer in the ODSMC to the uncertain discrete-time systems using an innovative LMI based framework.

The rest of this chapter is organized as follows: Section 3.2 describes the problem formulation. In Section 3.3, the proposed *scheme* to design an observer-based ODSMC with disturbance estimator is given. Effectiveness of the proposed ODSMC is shown by a numerical example in Section 3.4. Finally, Section 3.5 concludes this chapter.

3.2 Problem Formulation

Consider the following uncertain linear discrete-time system,

$$\begin{cases} x(k+1) = [A + \Delta A(k)]x(k) + B[u(k) + f(k)] \\ y(k) = Cx(k), \end{cases} \tag{3.1}$$

where $x(k) \in \mathbb{R}^n$, $u(k) \in \mathbb{R}^m$ and $y(k) \in \mathbb{R}^p$. Without loss of generality, it is assumed that $m \leq p \leq n$, $\text{rank}(B) = m$, $\text{rank}(C) = p$. Besides, it is assumed that (A,B) is controllable and (A,C) is observable. The uncertain matrix $\Delta A(k)$ is of the structure in (2.2).

In what follows, it is assumed that the exogenous disturbance in the system (3.1) is smooth and bounded.

Assumption 3.1 *The exogenous disturbance $f(k)$ in* (3.1) *satisfies the Lipschitz continuity condition,*

$$\|f_d(k)\| \leq L_f T_s, \quad \forall k \geq 0, \tag{3.2}$$

where $f_d(k) = f(k) - f(k-1)$, $L_f > 0$ denotes a Lipschitz constant and T_s is the sampling time.

Here, it is supposed that L_f has a small value. To this end, the sampling rate of the discrete signal processing system is assumed to be large enough compared to the maximum frequency component of the exogenous disturbance $f(k)$. Also, the following assumption is required to be considered in the sequel of this chapter.

Assumption 3.2 *The matrices A, B and C in the system* (3.1) *satisfy*

$$rank\left(\begin{bmatrix} A - I_n & B \\ C & 0 \end{bmatrix}\right) = n + m.$$

Notice that the above assumption requires that $m \leq p \leq n$, which has already been assumed in this chapter, and is equivalent to not having transmission zero at 1. Consider the following system state and disturbance observer

$$\begin{cases} \hat{x}(k+1) = A\hat{x}(k) + Bu(k) + L_1[y(k) - \hat{y}(k)] + B\hat{f}(k) \\ \hat{f}(k+1) = \hat{f}(k) + L_2[y(k) - \hat{y}(k)] \\ \hat{y}(k) = C\hat{x}(k), \end{cases} \tag{3.3}$$

where $L_1 \in \mathbb{R}^{n \times p}$ and $L_2 \in \mathbb{R}^{m \times p}$ are observer gains. The following lemma is useful in the sequel.

Lemma 3.1 ([64]) *For a given $B \in \mathbb{R}^{n \times m}$ with rank$(B) = m$, and*

$$B = U \begin{bmatrix} \Sigma \\ 0 \end{bmatrix} V^T, \tag{3.4}$$

where $U \in \mathbb{R}^{n \times n}$ and $V \in \mathbb{R}^{m \times m}$ are two orthogonal matrices and $\Sigma :=$

$diag(\sigma_1, \cdots, \sigma_m)$, $\sigma_i, (i = 1, \cdots, m)$ denote nonzero singular values of B, suppose that $0 < P \in \mathbb{R}^{n\times n}$ is a real symmetric matrix, then there exists a real matrix $Z \in \mathbb{R}^{m\times m}$ such that

$$PB = BZ, \tag{3.5}$$

if and only if P has the following structure

$$P = U \begin{bmatrix} P_{11} & 0 \\ 0 & P_{22} \end{bmatrix} U^T,$$

where $0 < P_{11} \in \mathbb{R}^{m\times m}$ and $0 < P_{22} \in \mathbb{R}^{(n-m)\times(n-m)}$.

In the rest of this chapter, for simplification, we use the brief ΔA instead of $\Delta A(k)$.

3.3 Observer-Based Output Feedback Discrete-Time SMC

In this section, the objective is to design a linear sliding function in the state space, such as

$$\sigma(k) = Sx(k), \tag{3.6}$$

where $S = B^T P_1 \in \mathbb{R}^{m\times n}$ and $P_1 > 0$ is a symmetric matrix that will be designed later. As seen, this structure of S would result in the non-singularity of SB. During the ideal sliding motion the sliding function satisfies

$$\sigma(k) = 0, \ \ \forall k > k_s, \tag{3.7}$$

where $k_s > 0$ denotes the time that sliding motion starts.

Remark 3.1 *In the case of CSMC, since the sliding function plays an important role in the discontinuous component of the controller, the switching function should be an entirely known one. Due to this fact, in the literature; e.g. [103], [147] and [35], the sliding function* (3.6) *has been supposed to satisfy*

$$B^T P_1 = GC, \tag{3.8}$$

in which $G \in \mathbb{R}^{m\times p}$. Then, the sliding surface (3.6) *can be rewritten as*

$$\sigma(k) = GCx(k) = Gy(k),$$

which is in the output space. However, since this switching function would not be used in the discrete-time sliding mode controller, such an equality as (3.8) *is unnecessary here. In fact, for the output feedback DSMC, the sliding surface is not required to be a known one, so, it will only need to be proved that system state trajectories could be steered into a boundary layer around the sliding surface and be kept there thereafter. The same manner can be seen in [83] for the static ODSMC.*

The controller is assumed to be of the following structure

$$u(k) = -(SB)^{-1}(SA - \Phi S)\hat{x}(k) - \hat{f}(k), \tag{3.9}$$

where $\Phi \in \mathbb{R}^{m\times m}$ is a stable matrix. The term $(SB)^{-1}\Phi S\hat{x}(k)$ would govern the rate of convergence onto the sliding manifold, in the absence of the external disturbance. Note that, unlike CSMC in which the so-called *equivalent controller* $-(SB)^{-1}SA\hat{x}(k)$ alone cannot steer the closed-loop system state trajectories on the ideal sliding surface, in the case of discrete-time systems the equivalent controller is able to drive the state trajectories of the discrete-time system into a neighborhood of the sliding manifolds and keeps them there thereafter [67]. However, with $\Phi = 0$ the control input aims at steering the system state to the sliding surface in one time step. In the case of a large initial distance from the sliding surface, this can lead to excessively large control input referred to as high-gain controller. Here, similar to [40], it is assumed that $\Phi = \lambda I_m$, where $0 \leq \lambda < 1$ is a given constant value which would not belong to the spectrum of A. Due to the special form of Φ, it can commute with S and then the control law (3.9) could be written as

$$u(k) = -(SB)^{-1}SA_{\lambda}\hat{x}(k) - \hat{f}(k), \tag{3.10}$$

where $A_{\lambda} = A - \lambda I_n$. Besides, we have

$$u_l(k) = -(SB)^{-1}SA_{\lambda}\hat{x}(k). \tag{3.11}$$

Defining the state estimation error $e(k) = x(k) - \hat{x}(k)$ and disturbance estimation error $e_f(k) = f(k) - \hat{f}(k)$, the overall closed-loop system is obtained by applying the controller in (3.10) to (3.1), which is

$$\begin{cases} x(k+1) = [A + \Delta A - \hat{A}]x(k) + B(SB)^{-1}S\left[\begin{smallmatrix} A_{\lambda} & B \end{smallmatrix}\right]e_a(k) \\ e_a(k+1) = \left[\begin{smallmatrix} \Delta A \\ 0 \end{smallmatrix}\right]x(k) + (A_a - L_aC_a)e_a(k) + \bar{f}_d(k+1), \end{cases} \tag{3.12}$$

where $\bar{f}_d(k+1) = \left[\begin{smallmatrix} 0 \\ f_d(k+1) \end{smallmatrix}\right]$, $e_a(k) = \left[\begin{smallmatrix} e(k) \\ e_f(k) \end{smallmatrix}\right]$, $A_a = \left[\begin{smallmatrix} A & B \\ 0 & I_m \end{smallmatrix}\right]$, $L_a = \left[\begin{smallmatrix} L_1 \\ L_2 \end{smallmatrix}\right]$ and $C_a = \left[\begin{smallmatrix} C & 0 \end{smallmatrix}\right]$. Then from (3.6) and (3.12) it can be found

$$\sigma(k+1) = \lambda\sigma(k) + S\Delta Ax(k) + S\left[\begin{smallmatrix} A_{\lambda} & B \end{smallmatrix}\right]e_a(k). \tag{3.13}$$

Lemma 3.2 ([32]) *If the matrix pair (A, C) is observable and A, B and C satisfy the rank condition in Assumption 3.2, then the matrix pair (A_a, C_a) is observable.*

Remark 3.2 *Note that exploiting the disturbance estimate in the ODSMC requires that the exogenous disturbances do not vary too much in one time step. This cannot only reduce the thickness of the boundary layer, but also relax the upper bound restriction on the exogenous disturbances, which can be seen in many references in the literature. Alternatively, this restriction is now on the maximum frequency component of the change of disturbance in terms of the sampling rate (see the continuity assumption in* (3.2)*).*

The sequel of this section aims to consider the boundedness of the system (3.1) using the controller (3.10).

3.3.1 Stability analysis

Notice that in the case of applying DSMC to the system involving exogenous disturbance, it can only ensure the state trajectories to be driven into a boundary layer around the ideal sliding surface $\sigma(k) = 0$. This issue is indeed regarded as the quasi sliding mode (QSM) in the literature. The following theorem considers a method to analyze simultaneously the reachabiltiy of QSM and the stability of the system states utilizing a discrete-time Lyapunov stability method, in the absence of exogenous disturbances. The characterization of the bounds on the closed-loop system states and sliding function's boundary layer are presented separately later in Theorem 3.2. Furthermore, as Theorem 3.2 needs to derive the cross terms between the system state (sliding function) and the component $\bar{f}_d(k+1)$, in order to avoid unnecessary repetition of the technical manipulations, we will start the proof of Theorem 3.1 more generally (with the external disturbance and the component $\bar{f}_d(k+1)$) for the sake of Theorem 3.2. We then let $f_d(k) = 0$ and thus $\bar{f}_d(k+1) = 0$ to derive the LMI condition for the stability analysis, and control and observer synthesis.

Theorem 3.1 *In the absence of $f(k)$, the control law* (3.10) *can drive the system state onto the ideal sliding surface $\sigma(k) = 0$, where $\sigma(k)$ is defined in* (3.6)*, and thus, the system state is stabilized if there exist symmetric matrices $P_1 := U\left[\begin{smallmatrix} P_{11} & 0 \\ 0 & P_{22}\end{smallmatrix}\right]U^T > 0$, $Q_2 > 0$, real matrices X_1, X_2 and X_3, and scalars $\varepsilon > 0$, $\rho > 0$ satisfying the following LMI:*

$$\begin{bmatrix} \mathscr{M}_{11} & \star & \star & \star & \star & \star & \star \\ 0 & -Q_2+\rho I & \star & \star & \star & \star & \star \\ \sqrt{2}\lambda B^T P_1 & \sqrt{2}B^T P_1 \left[\begin{smallmatrix} A_\lambda & B \end{smallmatrix}\right] & -B^T P_1 B & \star & \star & \star & \star \\ 0 & Q_2 A_a - X_3 C_a & 0 & -Q_2 & \star & \star & \star \\ P_1 A + BX_1 & 0 & 0 & 0 & -P_1 & \star & \star \\ BX_2 & 0 & 0 & 0 & 0 & -P_1 & \star \\ 0 & 0 & \sqrt{2}M^T P_1 B & \left[\begin{smallmatrix} M^T & 0 \end{smallmatrix}\right] Q_2 & M^T P_1 & 0 & -\varepsilon I \end{bmatrix} < 0, \tag{3.14}$$

where M and N are known matrices in (2.2)*, $0 < P_{11} \in \mathbb{R}^{m\times m}$, $0 < P_{22} \in \mathbb{R}^{(n-m)\times(n-m)}$, and $U \in \mathbb{R}^{n\times n}$ is defined in Lemma 2.5, $\mathscr{M}_{11} = -P_1 + X_2^T B^T + BX_2 + \rho I + \varepsilon N^T N$. Here $S = B^T P_1$ and the observer gains are given by*

$$\begin{bmatrix} L_1 \\ L_2 \end{bmatrix} = Q_2^{-1} X_3. \tag{3.15}$$

Proof 8 *Define*

$$V(\varpi(k)) = x^T(k)P_1 x(k) + e_a^T(k)Q_2 e_a(k) + \sigma^T(k)(SB)^{-1}\sigma(k), \tag{3.16}$$

where $\varpi(k) = \begin{bmatrix} x^T(k) & e_a^T(k) & \sigma^T(k)\end{bmatrix}^T$, $P_1 > 0$ and $Q_2 > 0$ are symmetric matrices and $S = B^T P_1$. Hence, we have

$$\begin{aligned} \Delta V(\varpi(k)) =& V(\varpi(k+1)) - V(\varpi(k)) \\ =& x^T(k+1)P_1 x(k+1) + e_a^T(k+1)Q_2 e_a(k+1) + \sigma^T(k+1)(SB)^{-1}\sigma(k+1) \\ & - x^T(k)P_1 x(k) - e_a^T(k)Q_2 e_a(k) - \sigma^T(k)(SB)^{-1}\sigma(k). \end{aligned} \tag{3.17}$$

It can follow then

$$
\begin{aligned}
& x^T(k+1)P_1x(k+1) \\
=& x^T(k)[A+\Delta A-B(SB)^{-1}S(A+\Delta A)+B(SB)^{-1}S(\lambda I_n+\Delta A)]^T \\
& \times P_1[A+\Delta A-B(SB)^{-1}S(A+\Delta A)+B(SB)^{-1}S(\lambda I_n+\Delta A)]x(k) \\
& +2x^T(k)(\lambda I_n+\Delta A)^T S^T(SB)^{-1}S\begin{bmatrix}A_\lambda & B\end{bmatrix}e_a(k) \\
& +e_a^T(k)\begin{bmatrix}A_\lambda & B\end{bmatrix}^T S^T(SB)^{-1}S\begin{bmatrix}A_\lambda & B\end{bmatrix}e_a(k) \\
=& x^T(k)[A+\Delta A-B(SB)^{-1}S(A+\Delta A)]^T P_1[A+\Delta A-B(SB)^{-1}S(A+\Delta A)]x(k) \\
& +x^T(k)(\lambda I_n+\Delta A)^T S^T(SB)^{-1}S(\lambda I_n+\Delta A)x(k)+2x^T(k)(\lambda I_n+\Delta A)^T S^T \\
& \times(SB)^{-1}S\begin{bmatrix}A_\lambda & B\end{bmatrix}e_a(k)+e_a^T(k)\begin{bmatrix}A_\lambda & B\end{bmatrix}^T S^T(SB)^{-1}S\begin{bmatrix}A_\lambda & B\end{bmatrix}e_a(k). \qquad (3.18)
\end{aligned}
$$

Also,

$$
\begin{aligned}
& e_a^T(k+1)Q_2e_a(k+1) \qquad (3.19) \\
=& x^T(k)\begin{bmatrix}\Delta A \\ 0\end{bmatrix}^T Q_2\begin{bmatrix}\Delta A \\ 0\end{bmatrix}x(k)+e_a^T(k)(A_a-L_aC_a)^T Q_2(A_a-L_aC_a)e_a(k) \\
& +\bar{f}_d^T(k+1)Q_2\bar{f}_d(k+1)+2x^T(k)\begin{bmatrix}\Delta A \\ 0\end{bmatrix}^T Q_2(A_a-L_aC_a)e_a(k) \\
& +2x^T(k)\begin{bmatrix}\Delta A \\ 0\end{bmatrix}^T Q_2\bar{f}_d(k+1)+2e_a^T(k)(A_a-L_aC_a)^T Q_2\bar{f}_d(k+1),
\end{aligned}
$$

$$
\begin{aligned}
& \sigma^T(k+1)(SB)^{-1}\sigma(k+1) \\
=& x^T(k)(\lambda I_n+\Delta A)^T S^T(SB)^{-1}S(\lambda I_n+\Delta A)x(k)+e_a^T(k)\begin{bmatrix}A_\lambda & B\end{bmatrix}^T S^T(SB)^{-1}S \\
& \times\begin{bmatrix}A_\lambda & B\end{bmatrix}e_a(k)+2x^T(k)(\lambda I_n+\Delta A)^T S^T(SB)^{-1}S\begin{bmatrix}A_\lambda & B\end{bmatrix}e_a(k). \qquad (3.20)
\end{aligned}
$$

From (3.17)-(3.20), *we have*

$$
\Delta V(\varpi(k))=\begin{bmatrix}x(k) \\ e_a(k) \\ \bar{f}_d(k+1)\end{bmatrix}^T\begin{bmatrix}\chi_{11} & \chi_{12} & \chi_{13} \\ \chi_{12}^T & \chi_{22} & \chi_{23} \\ \chi_{13}^T & \chi_{23}^T & \chi_{33}\end{bmatrix}\begin{bmatrix}x(k) \\ e_a(k) \\ \bar{f}_d(k+1)\end{bmatrix}, \qquad (3.21)
$$

where

$$\begin{aligned}
\chi_{11} =&(A+\Delta A)^T P_1 (A+\Delta A) - (A+\Delta A)^T S^T (SB)^{-1} S(A+\Delta A) \\
&- S^T (SB)^{-1} S - P_1 + 2(\lambda I_n + \Delta A)^T S^T (SB)^{-1} S(\lambda I_n + \Delta A) \\
&+ \begin{bmatrix} \Delta A \\ 0 \end{bmatrix}^T Q_2 \begin{bmatrix} \Delta A \\ 0 \end{bmatrix}, \\
\chi_{12} =&2(\lambda I_n + \Delta A)^T S^T (SB)^{-1} S \begin{bmatrix} A_\lambda & B \end{bmatrix} + \begin{bmatrix} \Delta A \\ 0 \end{bmatrix}^T Q_2 (A_a - L_a C_a), \\
\chi_{13} =& \begin{bmatrix} \Delta A \\ 0 \end{bmatrix}^T Q_2, \\
\chi_{22} =&2 \begin{bmatrix} A_\lambda & B \end{bmatrix}^T S^T (SB)^{-1} S \begin{bmatrix} A_\lambda & B \end{bmatrix} + (A_a - L_a C_a)^T Q_2 (A_a - L_a C_a) - Q_2, \\
\chi_{23} =&(A_a - L_a C_a)^T Q_2, \\
\chi_{33} =&Q_2.
\end{aligned}$$

Now, in order to analyze the system stability let $\bar{f}(k+1) = 0$. *Then the system is stabilized if*

$$\Upsilon := \begin{bmatrix} \chi_{11} & \chi_{12} \\ \chi_{12}^T & \chi_{22} \end{bmatrix} < -\rho I, \tag{3.22}$$

where $\rho > 0$ *is a scalar variable. Now it remains to consider the feasibility of* $\Upsilon < -\rho I$ *in* (3.22). *With the aid of the Schur complement,* (3.22) *is equivalent to*

$$\begin{bmatrix} \bar{\chi}_{11} & \star & \star & \star \\ 0 & \bar{\chi}_{22} & \star & \star \\ \sqrt{2}B^T P_1 (\lambda I_n + \Delta A) & \sqrt{2}B^T P_1 \begin{bmatrix} A_\lambda & B \end{bmatrix} & -B^T P_1 B & \star \\ Q_2 \begin{bmatrix} \Delta A \\ 0 \end{bmatrix} & Q_2 (A_a - L_a C_a) & 0 & -Q_2 \end{bmatrix} < 0, \tag{3.23}$$

where

$$\begin{aligned}
\bar{\chi}_{11} =&(A+\Delta A)^T P_1 (A+\Delta A) - (A+\Delta A)^T S^T (SB)^{-1} S(A+\Delta A) \\
&- S^T (SB)^{-1} S - P_1 + \rho I, \\
\bar{\chi}_{22} =& - Q_2 + \rho I.
\end{aligned}$$

Therefore, using Lemma 2.3 it can be shown that the feasibility of the inequality in (3.23) *is equivalent to that of*

$$\begin{bmatrix} \hat{\chi}_{11} & \star & \star & \star \\ 0 & \bar{\chi}_{22} & \star & \star \\ \sqrt{2}B^T P_1 (\lambda I_n + \Delta A) & \sqrt{2}B^T P_1 \begin{bmatrix} A_\lambda & B \end{bmatrix} & -B^T P_1 B & \star \\ Q_2 \begin{bmatrix} \Delta A \\ 0 \end{bmatrix} & Q_2 (A_a - L_a C_a) & 0 & -Q_2 \end{bmatrix} < 0, \tag{3.24}$$

with

$$\begin{aligned}
\hat{\chi}_{11} =&(A+\Delta A + BF_3)^T P_1 (A+\Delta A + BF_3) - P_1 \\
&+ F_4^T (B^T P_1 B) F_4 + F_4^T B^T P_1 + P_1 B F_4 + \rho I,
\end{aligned} \tag{3.25}$$

where F_3 and F_4 are two auxiliary variables [86]. Hence, using Lemma 3.1, $\hat{\chi}_{11}$ in (3.25) *can be rearranged as*

$$\hat{\chi}_{11} = [P_1(A+\Delta A)+BZF_3]^T P_1^{-1}[P_1(A+\Delta A)+BZF_3]-P_1 \\ +F_4^T Z^T B^T P_1^{-1} BZF_4 + F_4^T Z^T B^T + BZF_4 + \rho I, \tag{3.26}$$

where Z satisfies $P_1 B = BZ$. Using the Schur complement it can be shown that (3.24) *is equivalent to*

$$\begin{bmatrix} \tilde{\mathcal{M}}_{11} & \star & \star & \star & \star & \star \\ 0 & \bar{\chi}_{22} & \star & \star & \star & \star \\ \sqrt{2}B^T P_1(\lambda I_n + \Delta A) & \sqrt{2}B^T P_1 \left[A_\lambda \;\; B \right] & -B^T P_1 B & \star & \star & \star \\ Q_2 \left[\begin{smallmatrix} \Delta A \\ 0 \end{smallmatrix} \right] & Q_2 A_a - X_3 C_a & 0 & -Q_2 & \star & \star \\ P_1(A+\Delta A)+BX_1 & 0 & 0 & 0 & -P_1 & \star \\ BX_2 & 0 & 0 & 0 & 0 & -P_1 \end{bmatrix} < 0, \tag{3.27}$$

where $\tilde{\mathcal{M}}_{11} = -P_1 + X_2^T B^T + BX_2 + \rho I$, $X_1 = ZF_3$, $X_2 = ZF_4$ and $X_3 = Q_2 L_a$. With the help of Lemma 2.1 and the Schur complement, (3.27) *is sufficed by the LMI in* (3.14).

Remark 3.3 *The proof of this theorem provides a less conservative sufficient condition for the design of a robust sliding matrix for the system in* (3.1) *involving mismatched uncertainties. Further based on this proof, the second objective of this chapter, when the disturbance estimator is utilized in the controller directly, will be derived in the proof of Theorem 3.2.*

3.3.2 Characterizing the system state boundedness

Theorem 3.1 presents a framework to design the ODSMC in order to stabilize the system in (3.1). However, in the presence of exogenous disturbances, the proposed control law in (3.10) can only ensure the boundedness of the system state and sliding function. The following theorem characterizes the boundedness of the obtained closed-loop system state and corresponding sliding function.

Theorem 3.2 *In the presence of disturbance $f(k)$, if the LMI in* (3.14) *is feasible, for the obtained P_1, Q_2, $L_a = Q_2^{-1}X_3$ and ρ, the controller* (3.10) *satisfying* (3.2) *will lead to a bound on the augmented system state $\zeta(k) = [x^T(k), e_a(k), \sigma^T(k)]^T$ as follows:*

$$\forall \varepsilon > 0, \; \exists k^\star > 0, \; s.t. \; \forall k > k^\star, \\ \|\varpi(k)\|^2 \leq \frac{\lambda_{\max}(\mathbf{W})}{\hat{\rho}\lambda_{\min}(diag(P_1, Q_2, (B^T P_1 B)^{-1}))}\gamma + \varepsilon. \tag{3.28}$$

where $\mathbf{W} = \left[\begin{smallmatrix} M_{P_1} & 0 \\ 0 & Q_2 \end{smallmatrix} \right]$, with $M_{P_1} = P_1 B(B^T P_1 B)^{-1} B^T P_1 + P_1$, and $\gamma = \|\Pi + Q_2\| L_f^2 T_s^2$; here the scalar variable $\hat{\rho} > 0$ and matrix variable $\Pi > 0$ are obtained from solving *the following LMI:*

$$\begin{bmatrix} (\hat{\rho}-\rho)I+\tilde{\varepsilon}N^TN & \star & \star & \star \\ 0 & (\hat{\rho}-\rho)I & \star & \star \\ 0 & Q_2A_a - X_3C_a & -\Pi & \star \\ 0 & 0 & \begin{bmatrix} M^T & 0 \end{bmatrix} Q_2 & -\tilde{\varepsilon}I \end{bmatrix} < 0, \tag{3.29}$$

where $\tilde{\varepsilon} > 0$ is a scalar variable.

Proof 9 *Defining $\bar{x}(k) = \begin{bmatrix} x^T(k) & e_a^T(k) \end{bmatrix}^T$ and using Lemma 2.2 it can be shown that*

$$\begin{aligned} 2\bar{x}^T(k)\begin{bmatrix} \chi_{13} \\ \chi_{23} \end{bmatrix}\bar{f}_d(k+1) \leq & \bar{x}^T(k)\begin{bmatrix} \chi_{13} \\ \chi_{23} \end{bmatrix}\Pi^{-1}\begin{bmatrix} \chi_{13} \\ \chi_{23} \end{bmatrix}^T\bar{x}(k) \\ & + \bar{f}_d^T(k+1)\Pi\bar{f}_d(k+1), \end{aligned} \tag{3.30}$$

where $\Pi > 0$. It then follows from (3.21) *and* (3.30) *that*

$$\begin{aligned} \Delta V(\varpi(k)) \leq & -\bar{x}^T(k)\left\{\rho I - \begin{bmatrix} \chi_{13} \\ \chi_{23} \end{bmatrix}\Pi^{-1}\begin{bmatrix} \chi_{13} \\ \chi_{23} \end{bmatrix}^T\right\}\bar{x}(k) \\ & + \bar{f}_d^T(k+1)[\Pi + \chi_{33}]\bar{f}_d(k+1). \end{aligned} \tag{3.31}$$

Choosing $\Pi > 0$ subject to

$$\hat{\rho}I < \rho I - \begin{bmatrix} \chi_{13} \\ \chi_{23} \end{bmatrix}\Pi^{-1}\begin{bmatrix} \chi_{13} \\ \chi_{23} \end{bmatrix}^T, \tag{3.32}$$

where $0 < \hat{\rho} < \rho$, which is clearly always possible if $\eta > 0$ exists, it follows from (3.31) *that*

$$\Delta V(\varpi(k)) \leq -\hat{\rho}\bar{x}^T(k)\bar{x}(k) + \bar{f}_d^T(k+1)[\Pi + \chi_{33}]\bar{f}_d(k+1). \tag{3.33}$$

On the other hand, it can be seen that

$$\begin{aligned} V(\varpi(k)) = & \bar{x}^T(k)\begin{bmatrix} M_{P_1} & 0 \\ 0 & Q_2 \end{bmatrix}\bar{x}(k) \\ \triangleq & \bar{x}^T(k)\mathbf{W}\bar{x}(k), \end{aligned} \tag{3.34}$$

where $M_{P_1} = P_1B(B^TP_1B)^{-1}B^TP_1 + P_1$, then

$$\lambda_{\min}(\mathbf{W})\|\bar{x}(k)\|^2 \leq V(\varpi(k)) \leq \lambda_{\max}(\mathbf{W})\|\bar{x}(k)\|^2. \tag{3.35}$$

Furthermore, it can be shown that

$$\begin{aligned} \lambda_{\min}(diag(P_1, Q_2, (B^TP_1B)^{-1}))\|\varpi(k)\|^2 \leq & V(\varpi(k)) \\ \leq & \lambda_{\max}(diag(P_1, Q_2, (B^TP_1B)^{-1}))\|\varpi(k)\|^2. \end{aligned}$$

Hence, from (3.33) *and* (3.35)*, also the continuity assumption in* (3.2)*, we have*

$$\Delta V(\varpi(k)) \leq -\frac{\hat{\rho}}{\lambda_{\max}(\mathbf{W})} V(\varpi(k)) + \gamma. \tag{3.36}$$

where $\gamma = \|\Pi + Q_2\| L_f^2 T_s^2$. *Note that from* (3.22) *it is known that*

$$\bar{x}^T(k)\Upsilon\bar{x}(k) = V(\varpi(k+1))\big|_{\bar{f}_d(k+1)=0} - V(\varpi(k)) < -\rho\bar{x}^T(k)\bar{x}(k). \tag{3.37}$$

It is obvious that $V(\varpi(k+1))\big|_{\bar{f}_d(k+1)=0} \geq 0$*, and thus, from* (3.37) *and* (3.35)*, we have* $\rho < \lambda_{\max}(\mathbf{W})$*. Therefore,* $\frac{\hat{\rho}}{\lambda_{\max}(\mathbf{W})} < 1$*. Eventually, from Lemma 2.4 and* (3.36)*, the bound in* (3.28) *can be obtained.*

Moreover, to find $\Pi > 0$ *in* (3.32)*, for given* $P_1 > 0$*,* $Q_2 > 0$*,* L_a *and* $\rho > 0$*, by utilizing Lemma 2.1 and the Schur complement,* (3.32) *is sufficed by* (3.29)*, in which* $\tilde{\varepsilon} > 0$ *is a scalar variable.*

3.3.3 Discussions

1) The solution of the LMI in (3.29) does not have direct influence on the controller design and the actual ultimate bound on the system state and/or sliding function, however, these parameters would lead us to determine a more accurate bound. Therefore, to obtain the minimum value of the bound in (3.28) the LMIs in (3.14) and (3.29) could be solved subject to a specific criterion. This issue is beyond the scope of this book and remains for future works.

2) Due to the full column rank of B, the columns of matrices B and P_1B are linearly independent if $P_1 > 0$. Consequently, if (3.5) holds for $P_1 > 0$ and Z, we have

$$rank(Z) \geq rank(BZ) = rank(P_1B) \geq rank(B) = m, \tag{3.38}$$

which clearly denotes the non-singularity of Z. Also, it can easily be shown that

$$Z^{-1} = V\Sigma^{-1}P_{11}^{-1}\Sigma V^T. \tag{3.39}$$

3) Furthermore, unlike [101], [103] and [147] which use Lemma 2.2 to eliminate the cross terms between the system state (state estimate), the estimation error and even disturbance which obviously imposes some conservatism on the problem, here instead, it has been shown that the mentioned cross terms would not influence the feasibility region of the final LMI condition. Moreover, this chapter, unlike [101] which uses Lemma 2.2 to deal with the negative terms in $\Delta V(\zeta(k))$ to make a convex problem, exploits Lemma 2.3 which is clearly a *lossless* technique and imposes no additional conservatism on the LMI condition.

4) It should be noticed that the parameter $\Phi = \lambda I$, $0 < \lambda < 1$ plays a significant role in the magnitude of the thickness of the boundary layer around the sliding surface [67]. From (3.13) it can be shown that

$$\sigma_i(k) = \lambda^k \sigma_i(0) + \sum_{j=0}^{k-1} \lambda^{k-1-j} \mathscr{D}_i(j), \; i = 1, \cdots, m, \tag{3.40}$$

where $\mathscr{D}(j) = S\Delta Ax(j) + SA_\lambda e(j) + SBe_f(j)$. Supposing $\bar{\mathscr{D}}_i = \max(\mathscr{D}_i(k))$, it follows then from (3.40),

$$\forall \varepsilon_i > 0,\ \exists k^\dagger > 0,\ \text{s.t.}\ \forall k > k^\dagger,\ \sigma_i(k) < \frac{1}{1-\lambda}\bar{\mathscr{D}}_i + \varepsilon_i,\ i = 1, \cdots, m. \tag{3.41}$$

Assuming $\gamma_{\sigma,i} \triangleq \frac{1}{1-\lambda}\bar{\mathscr{D}}_i + \varepsilon_i$, the boundary layer is

$$\gamma_\sigma = \sqrt{\sum_{i=1}^{m} \gamma_{\sigma,i}^2}. \tag{3.42}$$

As seen, the smallest boundary layer could be obtained by setting λ to zero. In that case, the discrete-time sliding mode controller steers the system state into the *quasi sliding mode band* only in one time step. As mentioned, this would result in a high-gain or excessively large control input which is not desirable for most of the practical systems since it can saturate the actuators of the control system. Hence, there is a tradeoff to be considered between the level of the control input and the thickness of the boundary layer.

5) The sliding surface in this scheme is set to be in the state space, this matter is significantly different from the sliding surface in [103] and [147] which is in the estimation error space or the state estimate space. The Lyapunov functional candidate also, in these references, contains the state estimate and the state estimation error. Here, instead we have used the system state directly in addition to the state estimation error and sliding function in the Lyapunov functional candidate. Roughly speaking, the main drawback of the schemes, given in [103] and [147], comes from the fact that in order to formulate an LMI problem, it is inevitable to use same positive definite decision variable P for both quadratic terms $x^T(k)Px(k)$ and $e^T(k)Pe(k)$; otherwise, a BMI problem can arise, which is not easy to handle. For example, [35] utilizes two different positive definite decision variables in its Lyapunov-based scheme for the design of a dynamic output feedback CSMC (OCSMC), which naturally leads to a BMI problem. Note that, as mentioned earlier, since a variable structure discontinuous controller is not provided for the proposed ODSMC by the means of the sliding function, the introduced sliding function here, can be defined to be in the state space. Furthermore, in this case we do not need to struggle with a BMI problem.

3.4 Simulation Results

Again we consider the double integrator system in (1.1), (1.3). Furthermore, we assume that only state x_1 is available, i.e. output matrix $C = \begin{bmatrix} 1 & 0 \end{bmatrix}$. Now, with a sampling time of 0.5 s, a discrete-time system is derived as

$$A = \begin{bmatrix} 1.0000 & 0.5000 \\ 0 & 1.0000 \end{bmatrix}, B = \begin{bmatrix} 0.2500 \\ 0.500 \end{bmatrix} \tag{3.43}$$

We also consider the following uncertainty parameters and disturbance in the system

$$M = \begin{bmatrix} 0.0500 & 0.1500 \end{bmatrix}^T, N = \begin{bmatrix} -0.0500 & 0.1000 \end{bmatrix},$$
$$R(k) = 0.3\sin(k).$$

Suppose

$$f(k) = -0.01\sin\left(\frac{k}{5}\right).$$

Solving the LMI (3.14), the following results are obtained:

$$P_1 = \begin{bmatrix} 3.8195 & -1.2674 \\ -1.2674 & 5.7206 \end{bmatrix}, Q_2 = \begin{bmatrix} 48.2080 & -23.8747 & -21.7125 \\ -23.8747 & 31.3124 & -2.6942 \\ -21.7125 & -2.6942 & 52.8123 \end{bmatrix},$$
$$L_1 = \begin{bmatrix} 1.6957 \\ 1.4650 \end{bmatrix}, L_2 = 0.5991, S = \begin{bmatrix} 0.3212 & 2.5435 \end{bmatrix},$$
$$\rho = 0.3412, \varepsilon = 30.7173.$$

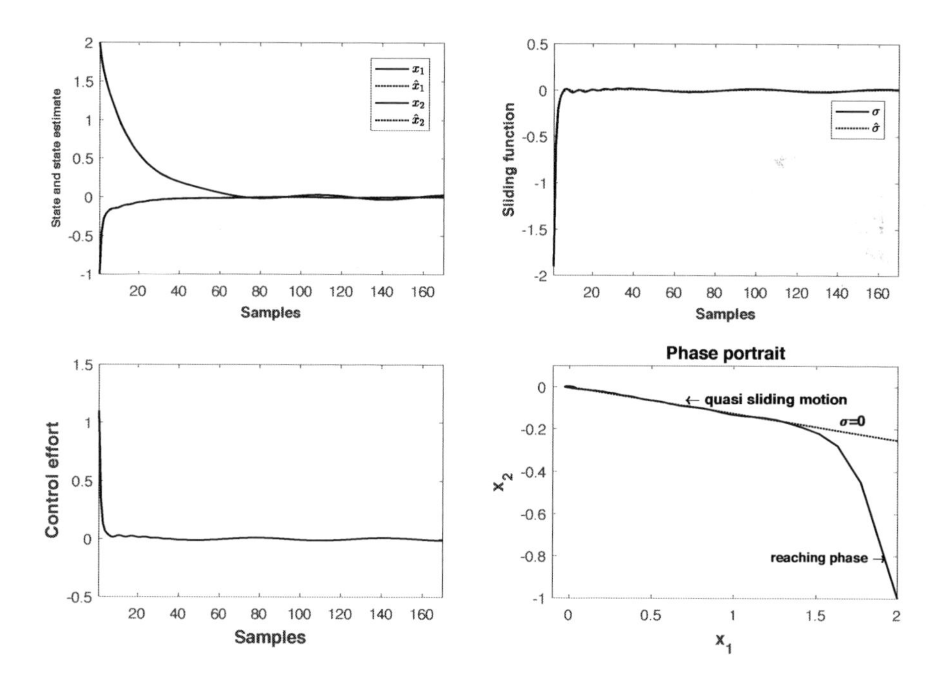

FIGURE 3.1
Results of ODSMC with disturbance estimator (3.10).

Applying the controller in (5.4) with given P_1 above to the system, the results are given in Fig. 3.1.

A comparison: It can be seen in the literature that a frequently used alternative for the discontinuous component (say $\vartheta(k)$) has the following general form:

$$\vartheta(k) = \varsigma + \upsilon \operatorname{sgn}(\sigma(k)), \tag{3.44}$$

where ς and υ are known parameters. For instance, in [101], with ignoring the bounds of $S\Delta Ax(k)$, ς and υ are set to be some constants which include the bounds of $SBf(k)$, similar to $\mathscr{F}^+$ and $\mathscr{F}^-$. Furthermore, due to the dependency of $\sigma(k)$ in (3.6) on the state $x(k)$, $\sigma(k)$ can also be replaced by

$$\hat{\sigma}(k) = S\hat{x}(k) = Sx(k) - Se(k),$$

and hence (2.19) is revised to

$$\vartheta(k) = \varsigma + \upsilon \operatorname{sgn}(\hat{\sigma}(k)). \tag{3.45}$$

Note that an LMI condition can be derived, as the one presented in Theorem 3.3 hereafter, that ensures the boundedness of $x(k)$ and $e(k)$, and hence, $\hat{\sigma}(k)$ can be an acceptable estimate of $\sigma(k)$. The controller containing $\vartheta(k)$ as in (3.45) may lead state trajectories to chatter around the switching surface with amplitude dependent on the lower bound on the term (3.45) and with the frequency equal to the sampling rate; see [67].

Let

$$\vartheta(k) = \mathscr{F}^+ + \varphi \operatorname{diag}(\mathscr{F}^-) \operatorname{sgn}(\hat{\sigma}(k)). \tag{3.46}$$

where $\varphi > 0$ is a designing parameter, $\mathscr{F}^+$ and $\mathscr{F}^-$ are defined in (2.11). Thus, the controller can be defined as:

$$u(k) = -(SB)^{-1} SA_\lambda \hat{x}(k) - \mathscr{F}^+ - \varphi \operatorname{diag}(\mathscr{F}^-) \operatorname{sgn}(\hat{\sigma}(k)), \tag{3.47}$$

Notice that the controller (3.47) satisfies all three reaching conditions given in [51]. Furthermore, in such a case, the observer in (3.3) can be replaced by a standard state observer as:

$$\begin{cases} \hat{x}(k+1) = A\hat{x}(k) + Bu(k) + L[y(k) - \hat{y}(k)] \\ \hat{y}(k) = C\hat{x}(k), \end{cases} \tag{3.48}$$

where $L \in \mathbb{R}^{n \times p}$ denotes the observer gain. The switching function matrix S and observer gain L will be selected through the following theorem.

Theorem 3.3 *The control law* (3.47) *can drive the system state into a boundary layer around the ideal sliding surface* (3.6) *and, in addition, the system state is ultimately bounded if there exist matrices* $P_1 := U \begin{bmatrix} P_{11} & 0 \\ 0 & P_{22} \end{bmatrix} U^T > 0$, $P_2 > 0$, X_1, X_2 *and* X_3, *and scalars* $\varepsilon > 0$, $\eta > 0$ *satisfying the following LMI:*

$$\begin{bmatrix} \mathscr{N}_{11} & \star & \star & \star & \star & \star & \star \\ 0 & -P_2 + \eta I & \star & \star & \star & \star & \star \\ \sqrt{2}\lambda B^T P_1 & \sqrt{2} B^T P_1 A_\lambda & -B^T P_1 B & \star & \star & \star & \star \\ 0 & P_2 A - X_3 C & 0 & -P_2 & \star & \star & \star \\ P_1 A + B X_1 & 0 & 0 & 0 & -P_1 & \star & \star \\ B X_2 & 0 & 0 & 0 & 0 & -P_1 & \star \\ 0 & 0 & \sqrt{2} M^T P_1 B & M^T P_2 & M^T P_1 & 0 & -\varepsilon I \end{bmatrix} < 0 \tag{3.49}$$

where M and N are known matrices in (2.2), $0 < P_{11} \in \mathbb{R}^{m\times m}$, $0 < P_{22} \in \mathbb{R}^{(n-m)\times(n-m)}$, *and* $U \in \mathbb{R}^{n\times n}$ *is defined in Lemma 3.1,* $\mathscr{N}_{11} = -P_1 + X_2^T B^T + BX_2 + \eta I + \varepsilon N^T N$. *Here* $S = B^T P_1$ *and the observer gain is given by*

$$L = P_2^{-1} X_3. \tag{3.50}$$

Proof 10 *The proof of this theorem is an application of the proof of Theorem 3.1 and thus is omitted here for the brevity purposes.*

Using the matrix S and the observer gain L obtained from solving the LMI condition (3.49), with $\varphi = 1$ and $\mathscr{F}^- = 0.01$, the results of applying the controller/observer in (3.47) and (3.48) to the system are given in Fig. 3.2.

As seen, the controller in (3.47) leads the state trajectories to chatter around the

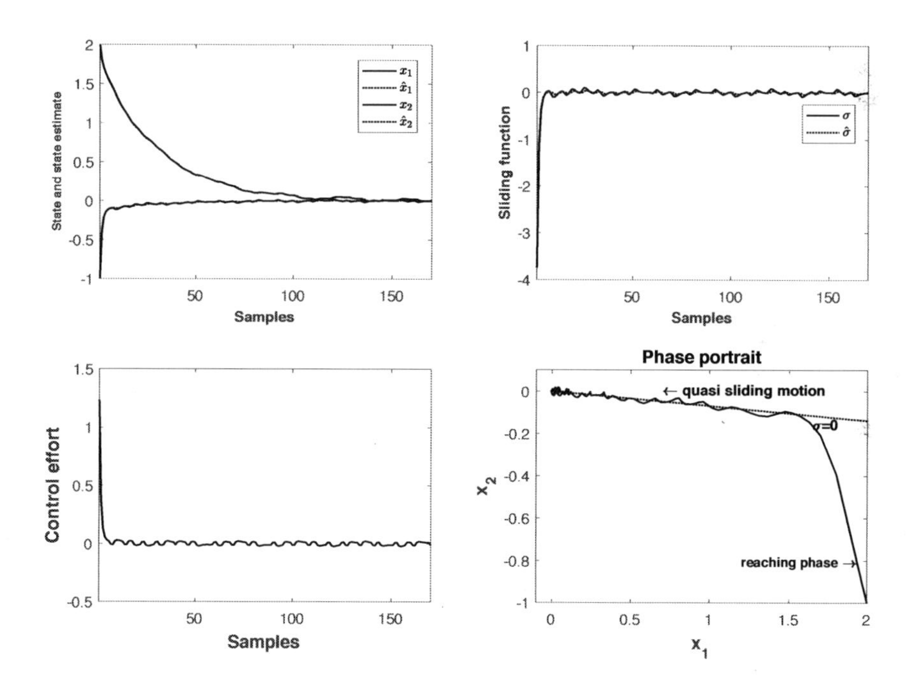

FIGURE 3.2
Results of applying ODSMC in (3.47).

switching surface, while the controller (3.10) that exploits the disturbance observer in its control law does not result in such a large and undesired chattering effect.

3.5 Conclusions

In this chapter, with the assumption of dealing with slow exogenous disturbances, a unified observer-based output feedback DSMC scheme, for the systems involving unmatched uncertainties and matched disturbances, has been developed, which includes an additional proportional integral estimator for estimating the disturbance. The proposed scheme is applicable to general systems including unstable systems. The boundedness of the obtained closed-loop system has been analyzed and a bound has been derived for the closed-loop system state, estimation error and also sliding function. The framework presented in this chapter is less conservative compared to the existing literature for the robust DSMC and also OCSMC.

Part II

DSMC for NCSs Involving Packet Losses

4

DSMC for NCSs involving consecutive measurement packet losses

CONTENTS

4.1	Introduction	69
4.2	Problem Formulation and Preliminaries	71
4.3	Stochastic Sliding Mode Control	73
4.4	Variable Structure Controller Considerations	80
4.5	Simulation Results	81
4.6	Conclusions	86

Abstract– This chapter develops a stabilizing sliding mode control for systems involving uncertainties as well as measurement data packet dropouts. In contrast to the existing literature that designs the switching function by using unavailable system states, a novel linear sliding function is constructed by employing only the available communicated system states for the systems involving measurement packet losses. This also equips us with the possibility to build a novel switching component for discrete-time sliding mode control DSMC by using only available system states. Finally, a numerical example is given to evaluate the performance of the designed DSMC for networked systems.

4.1 Introduction

Feedback control systems whose control loops involve real-time network are referred to as *networked control systems* (NCSs). The most significant advantages of NCSs, compared to the conventional control systems, are a smaller wiring system, lower overall cost, high reliability, as well as simple installation and maintenance etc; see [151]. However, using the communication network in the feedback control loop makes the stability analysis and controller design more complex. This fact is due to the possible delays and data packet dropouts which exist in the communication network (arising from its limited bandwidth) [146]. Hence, the problem of controlling the networked systems involving random delays and/or data packet dropout has been a subject of interest among researchers in the recent decades; e.g.

see [151, 146, 56, 91, 142, 154, 107, 49].

On the other hand, several probabilistic methods have been used to model the packet dropout so far (see [80, 100, 128]). Due to the practicality and simplicity of the *binary random delay*, this model has received more attention for modeling the network packet losses [141, 80, 100, 128, 137, 113].

A vast variety of the early discrete-time sliding mode control (DSMC) investigations have aimed to make a discrete-time counterpart to the continuous-time reachability condition [82, 117, 51]. Referring to the results presented in [67, 127, 98], the use of a switching function in the control law may not necessarily improve the performance. On the other hand, if the exogenous disturbance is bounded and smooth, it has been shown that exploiting a disturbance observer in the DSMC is beneficial in terms of reducing the thickness of the boundary layer around the sliding function [133]. Here, inspired by the idea of disturbance estimation, a novel form of switching function is proposed which can be more efficient in terms of reducing the ultimate bound on the system state and also bounding the chattering created by conventional switching functions.

Most of the work on SMC has been implemented for the systems in which system signals are transmitted perfectly; see e.g. [153, 83, 40]. Recently, several papers focused on the design of SMC for the systems involving time delays or packet losses. For instance, [152] develops an integral SMC in continuous time for offshore steel jacket platforms involving state time delays. Moreover, designing the DSMC subject to packet losses is considered in [101, 33]. However, the DSMC given in the existing literature suffers from the following major drawbacks:

- It is assumed that the packet dropout may not occur consecutively, which is obviously not a realistic assumption; *cf.* [101].
- The sliding function in the current literature is designed by means of the system state which, due to the packet losses in the communication channel, is not accessible; *cf.* [101, 33].
- The given methods utilize several inequalities, to provide an LMI condition for analyzing the boundedness of the overall closed-loop system and to design the sliding manifold. This imposes a heavy conservatism on the problem. Indeed, it requires that the open-loop system in [101] be stable.

The main goal of this chapter is to stabilize an NCS involving consecutive data packet dropout with a sliding mode control strategy. In doing so, a novel sliding function is introduced by employing the available communicated system states involving packet losses. This is significantly different from the existing DSMC in the literature [101, 33], and it also provides the possibility to directly build the switching component of the DSMC by exploiting only the available system states. In addition to this major progress, this chapter contains the following important innovations:

1. The proposed DSMC can be applied to the unstable systems directly *cf.* [101].

2. This chapter does not assume bounded system state in the first place to derive the controller or analyze the boundedness of the underlying closed-loop system. Also, the system uncertainty will be addressed using the robust control techniques; *cf.* [101, 27].
3. A novel scheme is developed to reduce the conservatism which exists in the current literature that removes the cross term between state and disturbance to make a fully diagonal problem.
4. In this chapter, in order to solve a matrix inequality problem including an uncertain negative signum quadratic term, a *lossless* technique is utilized to convert it to a form that can be easily written as an LMI. This technique can greatly widen the applicability region of our DSMC compared to the existing literature for the DSMC, *cf.* [101].
5. Considering smooth and bounded external disturbances, a novel stochastic disturbance observer is developed. A more practical switching function is provided in the controller using the proposed disturbance estimator with the aid of the *signum* function. The proposed DSMC that uses this switching function has effectively better performance in terms of reducing the thickness of the bound on the system state and sliding function in comparison to the linear controller.
6. This chapter utilizes a stochastic measurement model which can explain the occurrence of the long-term random delays and/or consecutive packet losses in the NCSs.

The rest of this chapter is organized as follows: Section 4.2 describes the problem formulation. In Section 4.3, the proposed method to design the stochastic sliding surface is given. Variable structure DSMC is discussed in Section 4.4. Efficiency of the proposed DSMC is studied by numerical examples in Section 4.5. Finally, Section 4.6 concludes this chapter.

4.2 Problem Formulation and Preliminaries

Consider the uncertain linear discrete-time system in (2.1). It is also assumed that the measured outputs, which are equivalent to the system states, are sent to the controller via a communication network. As a result, the measurements may involve the detrimental phenomenon referred to as *data packet dropout*. In this chapter, we use the following estimation scheme to provide the controller by the system state information,

$$x_c(k) = (1-\alpha(k))x(k) + \alpha(k)x_c(k-1), \tag{4.1}$$

where $x_c(k) \in \mathbb{R}^n$ is the communicated output available to the controller and the stochastic variable $\alpha(k) \in \mathbb{R}$ is a Bernoulli distributed white sequence with

$$\begin{aligned} \text{Prob}\{\alpha(k) = 1\} &= \mathbb{E}\{\alpha(k)\} = \bar{\alpha} \\ \text{Prob}\{\alpha(k) = 0\} &= 1 - \mathbb{E}\{\alpha(k)\} = 1 - \bar{\alpha}, \end{aligned} \tag{4.2}$$

where $0 \leq \bar{\alpha} < 1$ implies the probability that a data packet may be lost or has delay. Fig. 4.1 shows the structure of the NCS with (consecutive) random packet losses which will be considered in this chapter.

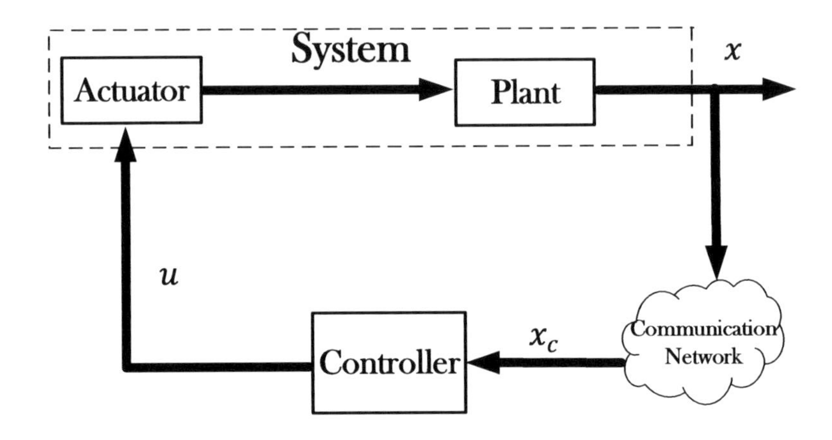

FIGURE 4.1
NCS structure.

Remark 4.1 *The measurement model* (4.1)*, proposed in [33], is essentially different from the one which has frequently been used in [113, 141, 146, 91, 101] etc. The estimator in* (4.1) *provides the controller with the previous existing system state estimate* $(x_c(k-1))$ *in the case of a packet dropout or delay. Note that in the literature, rather than the measurement model* (2.6)*, the following one is frequently utilized,*

$$x_c(k) = (1 - \alpha(k))x(k) + \alpha(k)x(k-1). \tag{4.3}$$

This model indeed would impose an assumption on the system networks that the packet dropout may not occur successively. This is obviously not a realistic assumption. Instead, the model (4.1) *can cope with the long-term random delays and/or frequent packet losses. Notice that the model* (4.3) *has been used to cope with both time varying communication delay (e.g. see [146, 91, 141]) and data packet dropout (e.g. see [101]). Considering the model* (2.6) *instead of model* (4.3)*, if a packet loss occurs, the estimator* (4.1) *provides the controller by* $x_c(k) = x_c(k-1)$*, or else the controller utilizes* $x_c(k) = x(k)$ *instead. Here,* $x_c(k-1)$ *denotes the last available communicated data packet. This justification can also be used for the communication random time-delay accordingly. It means that if the time-delay of the communication channel, which is assumed to be* τ_d*, is less than a sampling period* (T) *of the* *discrete-time control system, the delay has no influence on the system and we have* $x_c(k) = x(k)$*. But, if* $\tau_d \geq T$*, then* $x_c(k) = x_c(k-1)$*. Note that, in the case of random delay occurrence,* $x_c(k-1)$ *can be regarded as* $x(k-\tau_d)$*.*

In the rest of this chapter, for simplification, we use the brief α and ΔA instead of $\alpha(k)$ and $\Delta A(k)$, respectively. We also recall the notion of *exponential mean square stability* from [146] for the stochastic parameter systems as follows.

Definition 4.1 ([146]) *Consider the following stochastic system,*

$$\xi(k+1) = \Lambda\xi(k), \tag{4.4}$$

where $\xi(k) \in \mathbb{R}^n$ *denotes the state vector and* $\xi(0) \in \mathbb{R}^n$ *is its initial condition. If there exist constants* $\gamma > 0$ *and* $\rho \in (0,1)$ *such that*

$$\mathbb{E}\{\|\xi(k)\|^2\} \leq \gamma\rho^k \mathbb{E}\{\|\xi(0)\|^2\}, \quad \forall k > 0,$$

then the stochastic system in (4.4) *is said to be exponentially mean square stable.*

Lemma 4.1 ([146]) *Assume* $V(\xi(k))$ *to be a Lyapunov function. If there exist real scalars* $\lambda \geq 0$*,* $\mu > 0$*,* $v > 0$ *and* $0 < \psi < 1$ *such that*

$$\mu\,\|\xi(k)\|^2 \leq V(\xi(k)) \leq v\,\|\xi(k)\|^2,$$

and

$$\mathbb{E}\{V(\xi(k+1))|\xi(k)\} - V(\xi(k)) \leq \lambda - \psi V(\xi(k)), \tag{4.5}$$

then the sequence $\xi(k)$ *satisfies*

$$\mathbb{E}\{\|\xi(k)\|^2\} \leq \frac{v}{\mu}\mathbb{E}\{\|\xi(0)\|^2\}(1-\psi)^k + \frac{\lambda}{\mu\psi}.$$

Lemma 4.1 introduces a special boundedness definition for stochastic discrete-time systems involving exogenous disturbance. In this chapter, a system state which satisfies the condition presented in Lemma 2.4 is said to be exponentially mean square bounded.

4.3 Stochastic Sliding Mode Control

This chapter aims to design a stochastic sliding mode controller in order to stabilize the stochastic system (2.1).
Consider the following stochastic discrete-time sliding function,

$$\sigma_x(k) = Sx_c(k), \tag{4.6}$$

where the matrix $S \in \mathbb{R}^{m\times n}$ will be designed later such that SB is nonsingular. Notice that the sliding function used in this chapter is different from the one defined in [101]. This reference proposes an integral-like sliding surface by means of the system states $x(k)$ and $x(k-2)$. This sliding function could not be utilized directly in the variable structure discontinuous control strategy, as $x(k)$ and $x(k-2)$ are not available.

Notice that in the ideal sliding mode we have

$$\sigma_x(k+1) = \sigma_x(k) = 0, \ \ \forall k \geq k_s, \tag{4.7}$$

where $k_s > 0$ denotes the time when sliding motion starts. Now, the following control law is proposed,

$$u(k) = -(SB)^{-1}SAx_c(k) - \vartheta(k), \tag{4.8}$$

where $\varphi(k)$ denotes the nonlinear component which will be used in the controller to compensate for the bad effect of disturbance on the ultimate bound on the system state trajectories. It is assumed that the disturbance estimation component $\vartheta(k)$ is bounded, satisfying (2.13) (in the previous chapter). Some choices for $\vartheta(k)$ are presented later in Section 4.4.

Remark 4.2 *In this chapter, to construct the control law* (4.8)*, it is only assumed that the upper and lower bounds on the matched exogenous disturbance are known. However, it can be seen in the literature ([101] and [27]) that the term $S\Delta Ax(k)$ is assumed to be bounded. As $\Delta A(k)$ is a time-varying uncertainty, in order to find the bounds of $S\Delta Ax(k)$, one is required to know the behavior of the closed-loop system state in advance, and as a result, to know $\|x(k)\|$ from the beginning and before applying the controller to the system. Broadly speaking, it is not a realistic assumption that we can have the bounds on the system states. Instead, we neglect using any bound on the system states and only deal with the system unmatched (mismatched) uncertainties using the robust control strategies.*

The rest of this section aims to consider the stability of the stochastic system (2.1) using the controller (4.8). As a result of applying the controller (4.8) to the system (2.1), it is seen that

$$x(k+1) = \left[A + \Delta A - (1-\alpha)\hat{A}\right]x(k) - \alpha\hat{A}x_c(k-1) + Bf_\vartheta(k), \tag{4.9}$$

where $\hat{A} = B(SB)^{-1}SA$. Furthermore, it can be found that

$$\begin{aligned}\sigma_x(k+1) =&(1-\alpha_{k+1})[\alpha_k S(A+\Delta A) + (1-\alpha_k)S\Delta A]x(k)\\ &-\alpha_k(1-\alpha_{k+1})SAx_c(k-1) + (1-\alpha_{k+1})SBf_\vartheta(k)\\ &+\alpha_{k+1}(1-\alpha_k)Sx(k) + \alpha_{k+1}\alpha_k Sx_c(k-1),\end{aligned} \tag{4.10}$$

where α_{k+1} and α_k stand for $\alpha(k+1)$ and $\alpha(k)$ respectively. It should be noted that with applying DSMC to discrete-time systems involving exogenous disturbances, the closed-loop system should be analyzed in terms of boundedness. Besides, the DSMC would only ensure that the state trajectories are driven into a boundary layer around the ideal sliding surface $\sigma(k) = 0$. This issue is indeed regarded as the quasi sliding mode (QSM) in the literature. On the other hand, due to the presence of mismatched uncertainty in the system dynamics, it is difficult to analyze the reachability of the QSM by means of a separate sufficient condition. Alternatively, the following theorem considers a method to analyze simultaneously the reachabiltiy of QSM and the boundedness of the system states by means of a discrete-time Lyapunov stability method.

Theorem 4.1 *The control law* (4.8) *can steer the state of the stochastic system* (2.1) *into a boundary layer around the ideal sliding surface* (4.7) *and, also the system state is exponentially mean square bounded if there exist* $P := U\left[\begin{smallmatrix} P_{11} & 0 \\ 0 & P_{22} \end{smallmatrix}\right]U^T > 0$, $Q > 0$, X_i, $i = 1,2,3$, *and scalars* $\varepsilon > 0$, $\eta > 0$ *satisfying the following LMI,*

$$\begin{bmatrix} \check{\Psi}_{11} & \star & \star & \star & \star & \star & \star & \star \\ 0 & \check{\Psi}_{22} & \star & \star & \star & \star & \star & \star \\ \mu_1 B^T PA & -\mu_1 B^T PA & -B^T PB & \star & \star & \star & \star & \star \\ 0 & 0 & 0 & -B^T PB & \star & \star & \star & \star \\ BX_2 & 0 & 0 & 0 & -P & \star & \star & \star \\ PA + BX_1 & 0 & 0 & 0 & 0 & -P & \star & \star \\ 0 & BX_3 & 0 & 0 & 0 & 0 & -P & \star \\ 0 & 0 & \mu_1 M^T PB & \mu_2 M^T PB & 0 & M^T P & 0 & -\varepsilon I \end{bmatrix} < 0, \tag{4.11}$$

where $\check{\Psi}_{11} = -P + \bar{Q} + (1-\bar{\alpha})(X_2^T B^T + BX_2) + \eta I + \varepsilon N^T N$, $\check{\Psi}_{22} = -\bar{Q} + \delta(X_3^T B^T + BX_3) + \eta I$, $\mu_1 = \sqrt{\bar{\alpha}(2-\bar{\alpha})}$, $\mu_2 = \sqrt{(1-\bar{\alpha})(2-\bar{\alpha})}$, $\delta = \sqrt{\bar{\alpha}(1-\bar{\alpha})}$ *and* $\bar{Q} = (1-\bar{\alpha})Q$. *Here* U *is defined in* (3.4), $S = B^T P$ *and* $\{\star\}$ *denotes the symmetric elements in a symmetric matrix.*

Proof 11 *Define*

$$V(\zeta(k)) = x^T(k)Px(k) + x_c^T(k-1)Qx_c(k-1) + \sigma_x^T(k)(SB)^{-1}\sigma_x(k),$$

where $\zeta(k) = \begin{bmatrix} x^T(k) & x_c^T(k-1) & \sigma_x^T(k) \end{bmatrix}^T$, $P > 0$ *and* $Q > 0$ *are symmetric matrices and* $S = B^T P$. *Thus, we can write*

$$\begin{aligned} &\Delta V(\zeta(k)) \triangleq \mathbb{E}\{V(\zeta(k+1))|\zeta(k)\} - V(\zeta(k)) \\ =&\mathbb{E}\left\{x^T(k+1)Px(k+1) + x_c^T(k)Qx_c(k) + \sigma_x^T(k+1)(SB)^{-1}\sigma_x(k+1)|\zeta(k)\right\} \\ &- x^T(k)Px(k) - x_c^T(k-1)Qx_c(k-1) - \sigma_x^T(k)(SB)^{-1}\sigma_x(k). \end{aligned} \tag{4.12}$$

It can be shown then

$$\begin{aligned} &\mathbb{E}\{x^T(k+1)Px(k+1)|\zeta(k)\} \\ =&\mathbb{E}\Big\{x^T(k)\left[A + \Delta A - (1-\alpha)\hat{A}\right]^T P\left[A + \Delta A - (1-\alpha)\hat{A}\right]x(k) \\ &- 2x^T(k)\left[A + \Delta A - (1-\alpha)\hat{A}\right]^T P\alpha\hat{A}x_c(k-1) \\ &+ \alpha^2 x_c^T(k-1)A^T S^T(SB)^{-1}SAx_c(k-1) \\ &+ 2x^T(k)\left[A + \Delta A - (1-\alpha)\hat{A}\right]^T S^T f_\vartheta(k) \\ &- 2\alpha x_c^T(k-1)A^T S^T f_\vartheta(k) + f_\vartheta^T(k)(SB)f_\vartheta(k)\Big|\zeta(k)\Big\}. \end{aligned} \tag{4.13}$$

Besides, note that $\mathbb{E}\{\alpha(1-\alpha)\}=0$, $\mathbb{E}\{\alpha^2\}=\bar{\alpha}$ *and* $\mathbb{E}\{(1-\alpha)^2\}=1-\bar{\alpha}$. *Thus, it can be demonstrated that*

$$\begin{aligned}&\mathbb{E}\Big\{x^T(k)\big[A+\Delta A-(1-\alpha)\hat{A}_\Delta+(1-\alpha)\Delta\hat{A}\big]^T P\\&\times\big[A+\Delta A-(1-\alpha)\hat{A}_\Delta+(1-\alpha)\Delta\hat{A}\big]x(k)\Big|\zeta(k)\Big\}\\=&x^T(k)[(A+\Delta A)^T P(A+\Delta A)-(1-\bar{\alpha})(A+\Delta A)^T S^T(SB)^{-1}S(A+\Delta A)\\&+(1-\bar{\alpha})\Delta A^T S^T(SB)^{-1}S\Delta A]x(k),\end{aligned}\tag{4.14}$$

where $\hat{A}_\Delta=B(SB)^{-1}S(A+\Delta A)$, $\Delta\hat{A}=B(SB)^{-1}S\Delta A$ *and*

$$\begin{aligned}&\mathbb{E}\Big\{-2x^T(k)\big[A+\Delta A-(1-\alpha)\hat{A}\big]^T P\alpha\hat{A}x_c(k-1)\Big|\zeta(k)\Big\}\\=&-2\bar{\alpha}x^T(k)(A+\Delta A)^T S^T(SB)^{-1}SAx_c(k-1),\end{aligned}\tag{4.15}$$

$$\begin{aligned}&\mathbb{E}\Big\{\alpha^2x_c^T(k-1)A^TS^T(SB)^{-1}SAx_c(k-1)\Big|\zeta(k)\Big\}\\=&\bar{\alpha}x_c^T(k-1)A^TS^T(SB)^{-1}SAx_c(k-1),\end{aligned}\tag{4.16}$$

$$\begin{aligned}&\mathbb{E}\Big\{2x^T(k)\big[A+\Delta A-(1-\alpha)\hat{A}_\Delta+(1-\alpha)\Delta\hat{A}\big]^T S^T f_\vartheta(k)\Big|\zeta(k)\Big\}\\=&2x^T(k)(\bar{\alpha}A+\Delta A)^T S^T f_\vartheta(k),\end{aligned}\tag{4.17}$$

$$\begin{aligned}&\mathbb{E}\Big\{-2\alpha x_c^T(k-1)A^TS^Tf_\vartheta(k)\Big|\zeta(k)\Big\}\\=&-2\bar{\alpha}x_c^T(k-1)A^TS^Tf_\vartheta(k).\end{aligned}\tag{4.18}$$

Also, it is easy to show that

$$\begin{aligned}&\mathbb{E}\Big\{x_c^T(k)Qx_c(k)\Big|\zeta(k)\Big\}\\=&(1-\bar{\alpha})x^T(k)Qx(k)+\bar{\alpha}x_c^T(k-1)Qx_c(k-1),\end{aligned}\tag{4.19}$$

$$\begin{aligned}&\mathbb{E}\{\sigma_x^T(k+1)(SB)^{-1}\sigma_x(k+1)|\zeta(k)\}\\=&\delta^2x^T(k)(A+\Delta A)^TS^T(SB)^{-1}S(A+\Delta A)x(k)\\&+(1-\bar{\alpha})^2x^T(k)\Delta A^TS^T(SB)^{-1}S\Delta Ax(k)\\&-2\delta^2x^T(k)(A+\Delta A)^TS^T(SB)^{-1}SAx_c(k-1)+2\delta^2x^T(k)(A+\Delta A)^TS^Tf_\vartheta(k)\\&+2(1-\bar{\alpha})^2x^T(k)\Delta A^TS^Tf_\vartheta(k)+\delta^2x_c^T(k-1)A^TS^T(SB)^{-1}SAx_c(k-1)\\&-2\delta^2x_c^T(k-1)A^TS^Tf_\vartheta(k)+(1-\bar{\alpha})f_\vartheta^T(k)(SB)f_\vartheta(k)+\delta^2x^T(k)S^T(SB)^{-1}Sx(k)\\&+\bar{\alpha}^2x_c^T(k-1)S^T(SB)^{-1}Sx_c(k-1),\end{aligned}\tag{4.20}$$

$$\begin{aligned}&\mathbb{E}\{\sigma_x^T(k)(SB)^{-1}\sigma_x(k)|\zeta(k)\}\\&=(1-\bar{\alpha})x^T(k)S^T(SB)^{-1}Sx(k)+\bar{\alpha}x_c^T(k-1)S^T(SB)^{-1}Sx_c(k-1),\end{aligned} \tag{4.21}$$

in which $\mathbb{E}\{\alpha_{k+1}\alpha_k\}=\bar{\alpha}^2$, $\mathbb{E}\{(1-\alpha_{k+1})\alpha_k\}=\mathbb{E}\{(1-\alpha_k)\alpha_{k+1}\}=\bar{\alpha}(1-\bar{\alpha})\triangleq\delta^2$ *and* $\mathbb{E}\{(1-\alpha_k)(1-\alpha_{k+1})\}=(1-\bar{\alpha})^2$. *Using* (4.13)-(4.21)*, it follows from* (4.12) *that*

$$\Delta V(\zeta(k))=\begin{bmatrix}x(k)\\x_c(k-1)\\f_\vartheta(k)\end{bmatrix}^T\begin{bmatrix}\Psi_{11}&\Psi_{12}&\Psi_{13}\\\Psi_{12}^T&\Psi_{22}&\Psi_{23}\\\Psi_{13}^T&\Psi_{23}^T&\Psi_{33}\end{bmatrix}\begin{bmatrix}x(k)\\x_c(k-1)\\f_\vartheta(k)\end{bmatrix}, \tag{4.22}$$

where

$$\begin{aligned}\Psi_{11}=&(A+\Delta A)^TP(A+\Delta A)-(1-\bar{\alpha})^2(A+\Delta A)^TS^T(SB)^{-1}S(A+\Delta A)\\&+(1-\bar{\alpha})(2-\bar{\alpha})\Delta A^TS^T(SB)^{-1}S\Delta A-P+(1-\bar{\alpha})Q\\&-(1-\bar{\alpha})^2S^T(SB)^{-1}S,\\\Psi_{12}=&-\bar{\alpha}(2-\bar{\alpha})(A+\Delta A)^TS^T(SB)^{-1}SA,\\\Psi_{22}=&\bar{\alpha}(2-\bar{\alpha})A^TS^T(SB)^{-1}SA-\delta^2S^T(SB)^{-1}S-(1-\bar{\alpha})Q,\end{aligned}$$

and $\Psi_{13}=(2-\bar{\alpha})[\bar{\alpha}A+\Delta A]^TS^T$, $\Psi_{23}=-\bar{\alpha}(2-\bar{\alpha})A^TS^T$ *and* $\Psi_{33}=(2-\bar{\alpha})SB$. *Now, to prove the system stability, let* $f_\vartheta(k)=0$. *Then the system is stable if*

$$\Upsilon:=\begin{bmatrix}\Psi_{11}&\Psi_{12}\\\Psi_{12}^T&\Psi_{22}\end{bmatrix}<-\eta I, \tag{4.23}$$

where $\eta>0$ *is a scalar variable. With the choice of* $S=B^TP$ *and using Lemma 2.3, it is known that the feasibility of* (4.23) *is equivalent to that of*

$$\begin{bmatrix}\tilde{\Psi}_{11}&\star&\star&\star\\0&\tilde{\Psi}_{22}&\star&\star\\\mu_1B^TP(A+\Delta A)&-\mu_1B^TPA&-B^TPB&\star\\\mu_2B^TP\Delta A&0&0&-B^TPB\end{bmatrix}<0, \tag{4.24}$$

where $\mu_1=\sqrt{\bar{\alpha}(2-\bar{\alpha})}$, $\mu_2=\sqrt{(1-\bar{\alpha})(2-\bar{\alpha})}$, $\bar{Q}=(1-\bar{\alpha})Q$ *and*

$$\begin{aligned}\tilde{\Psi}_{11}=&(A+\Delta A+BF_1)^TP(A+\Delta A+BF_1)+F_2^T(B^TPB)F_2+(1-\bar{\alpha})F_2^TB^TP\\&+(1-\bar{\alpha})PBF_2-P+\bar{Q}+\eta I,\end{aligned} \tag{4.25}$$

$$\tilde{\Psi}_{22}=-\bar{Q}+\eta I+F_3^T(B^TPB)F_3+\delta F_3^TB^TP+\delta PBF_3, \tag{4.26}$$

in which $F_i, i=1,2,3$, *are introduced auxiliary variables as in Lemma 2.3. According to Lemma 3.1 and letting* $P:=U\left[\begin{smallmatrix}P_{11}&0\\0&P_{22}\end{smallmatrix}\right]U^T>0$, *where* $0<P_{11}\in\mathbb{R}^{m\times m}$ *and* $0<P_{22}\in\mathbb{R}^{(n-m)\times(n-m)}$, *there exists* $Z\in\mathbb{R}^{m\times m}$ *such that the equality in* (3.5) *holds.*

Then it follows from (4.25) *to have*

$$\tilde{\Psi}_{11} = [P(A+\Delta A) + BZF_1]^T P^{-1} [P(A+\Delta A) + BZF_1] + F_2^T Z^T B^T P^{-1} BZF_2 \\ + (1-\bar{\alpha}) F_2^T Z^T B^T + (1-\bar{\alpha}) BZF_2 - P + \bar{Q} + \eta I. \quad (4.27)$$

$$\tilde{\Psi}_{22} = -\bar{Q} + \eta I + F_3^T Z^T B^T P^{-1} BZF_3 + \delta F_3^T Z^T B^T + \delta BZF_3. \quad (4.28)$$

Using the Schur complement and Lemma 2.1, with introducing $ZF_i = X_i$, $i = 1,2,3$, *the inequality in* (4.24) *can be implied by the LMI* (4.11).

While the above theorem presents a framework to design the DSMC in order to stabilize the NCS in (2.1), it does not give a bound on the system states. The following theorem will characterize the boundedness of the obtained stochastic closed-loop system state and corresponding sliding function.

Theorem 4.2 *If the LMI in* (4.11) *is feasible then the bound on the augmented system state* $\zeta(k)$ *is, for given* $P > 0, Q > 0$ *and* $\eta > 0$, *as follows*

$$\forall \upsilon > 0,\ \exists k^\star > 0,\ \forall k > k^\star,\ s.t.\ \mathbb{E}\left\{\|\zeta(k)\|^2\right\} \leq \frac{\lambda_{\max}(\mathbf{M})}{\hat{\eta}\lambda_1}\gamma + \upsilon, \quad (4.29)$$

where $\lambda_1 = \lambda_{\min}[diag(P,Q,(SB)^{-1})]$, $\mathbf{M} = diag(M_p, Q)$ *with* $M_P = P + PB(B^T PB)^{-1} B^T P$, *and* $\gamma = \tau^2 \|\Pi + (2-\bar{\alpha})SB\| \, \|\mathscr{F}^{-}\|^2$; *here the scalar variable* $\hat{\eta} > 0$ *and matrix variable* $\Pi > 0$ *are obtained from solving the following LMI,*

$$\begin{bmatrix} (\hat{\eta}-\eta)I + \bar{\varepsilon} N^T N & \star & \star & \star \\ 0 & (\hat{\eta}-\eta)I & \star & \star \\ \mu_1^2 B^T PA & -\mu_1^2 B^T PA & -\Pi & \star \\ 0 & 0 & (2-\bar{\alpha})M^T PB & -\bar{\varepsilon} I \end{bmatrix} < 0, \quad (4.30)$$

where $\bar{\varepsilon} > 0$ *is a scalar variable.*

Proof 12 *Defining* $\bar{x}(k) = \begin{bmatrix} x^T(k) & x_c^T(k-1) \end{bmatrix}^T$ *and referring to Lemma 2.2 it can be written that*

$$2\bar{x}^T(k) \begin{bmatrix} \Psi_{13} \\ \Psi_{23} \end{bmatrix} f_\vartheta(k) \leq \bar{x}^T(k) \begin{bmatrix} \Psi_{13} \\ \Psi_{23} \end{bmatrix} \Pi^{-1} \begin{bmatrix} \Psi_{13} \\ \Psi_{23} \end{bmatrix}^T \bar{x}(k) + f_\vartheta^T(k) \Pi f_\vartheta(k), \quad (4.31)$$

where $\Pi > 0$ *is a matrix of appropriate dimension. It follows from* (4.22), (4.23) *and* (4.31) *that*

$$\Delta V(\zeta(k)) \leq -\bar{x}^T(k) \left\{ \eta I - \begin{bmatrix} \Psi_{13} \\ \Psi_{23} \end{bmatrix} \Pi^{-1} \begin{bmatrix} \Psi_{13} \\ \Psi_{23} \end{bmatrix}^T \right\} \bar{x}(k) + f_\vartheta^T(k)[\Pi + \Psi_{33}] f_\vartheta(k). \quad (4.32)$$

If we choose $\Pi > 0$ *such that*

$$\hat{\eta} I < \eta I - \begin{bmatrix} \Psi_{13} \\ \Psi_{23} \end{bmatrix} \Pi^{-1} \begin{bmatrix} \Psi_{13} \\ \Psi_{23} \end{bmatrix}^T, \quad (4.33)$$

where $0 < \hat{\eta} < \eta$, *which is always possible if* $\eta > 0$ *exists, then, it follows from* (4.32) *that*

$$\Delta V(\zeta(k)) \leq -\hat{\eta}\bar{x}^T(k)\bar{x}(k) + f_{\vartheta}^T(k)[\Pi + \Psi_{33}]f_{\vartheta}(k). \tag{4.34}$$

Besides, notice that

$$\begin{aligned} V(\zeta(k)) &= \bar{x}^T(k)\begin{bmatrix} M_P & 0 \\ 0 & Q \end{bmatrix}\bar{x}(k) \\ &\triangleq \bar{x}^T(k)\mathbf{M}\bar{x}(k), \end{aligned}$$

where $M_P = P + PB(B^TPB)^{-1}B^TP$, *hence*

$$\lambda_{\min}(\mathbf{M})\,\|\bar{x}(k)\|^2 \leq V(\zeta(k)) \leq \lambda_{\max}(\mathbf{M})\,\|\bar{x}(k)\|^2. \tag{4.35}$$

Moreover, it is obvious that

$$\lambda_1\,\|\zeta(k)\|^2 \leq V(\zeta(k)) \leq \lambda_2\,\|\zeta(k)\|^2, \tag{4.36}$$

where $\lambda_1 = \lambda_{\min}[diag(P, Q, (SB)^{-1})]$ *and* $\lambda_2 = \lambda_{\max}[diag(P, Q, (SB)^{-1})]$. *Therefore, from* (4.34) *and* (4.35), *one can derive that*

$$\Delta V(\zeta(k)) \leq -\frac{\hat{\eta}}{\lambda_{\max}(\mathbf{M})}V(\zeta(k)) + \gamma, \tag{4.37}$$

where

$$\gamma = \tau^2\,\|\Pi + (2-\bar{\alpha})SB\|\,\left\|\mathscr{F}^-\right\|^2. \tag{4.38}$$

Note that from (4.23) *it can simply be written that,* $\forall\bar{x}(k) \neq 0$

$$\begin{aligned} \mathbb{E}\left\{V(\zeta(k+1))\big|_{f_{\vartheta}(k)=0}\Big|\zeta(k)\right\} - V(\zeta(k)) &= \bar{x}^T(k)\Upsilon\bar{x}(k) \\ &< -\eta\bar{x}^T(k)\bar{x}(k). \end{aligned} \tag{4.39}$$

It is also known that $\mathbb{E}\left\{V(\zeta(k+1))\big|_{f_{\vartheta}(k)=0}\Big|\zeta(k)\right\} > 0$, *and thus, from* (4.39) *and* (4.35), *it can be claimed that* $\lambda_{\max}(\mathbf{M}) > \eta$. *Therefore*

$$\frac{\hat{\eta}}{\lambda_{\max}(\mathbf{M})} < 1.$$

Eventually, from Lemma 4.1 and (4.37) *one can reach the bound given in* (4.29). *Now to find* $\Pi > 0$ *and* $\hat{\eta}$ *utilized in* (4.29), *we need to check the feasibility of* (4.33). *We then use the Schur complement and Lemma 2.1, for given* $P > 0$, $Q > 0$ *and* $\eta > 0$, *to show that the inequality in* (4.33) *is sufficed by the LMI in* (4.30).

Remark 4.3 *Due to the full column rank of B, the columns of B and PB are linearly independent if* $P > 0$. *Consequently, if* (3.5) *holds for* $P > 0$ *and Z, we have*

$$rank(Z) \geq rank(BZ) = rank(PB) \geq rank(B) = m,$$

which clearly denotes the non-singularity of Z. Also, it can easily be shown that

$$Z^{-1} = V\Sigma^{-1}P_{11}^{-1}\Sigma V^T.$$

Remark 4.4 *The scheme exploited in the proof of Theorem 4.1 can reduce the conservatism of the sufficient conditions which can be found in the existing literature (cf. [101]). Indeed, unlike [101] which uses Lemma 2.2 to deal with the negative terms in* $\Delta V(\zeta(k))$ *to make a convex problem, we exploit Lemma 2.3 which is clearly a lossless technique and imposes no additional conservatism on the LMI condition. Also, in [101], the cross terms between the system state (*$x(k)$*,* $x_c(k-1)$*) and* $f_\vartheta(k)$ *have increased the conservatism of the final obtained LMI condition. Here, it has been shown that the mentioned cross terms would not influence the feasibility region of the final LMI condition* (4.11). *Additionally, [101] removes the cross term between* $x(k)$ *and* $x_c(k-1)$ *(*$x(k-1)$ *in [101]) through an inequality. This chapter avoids using this inequality and lets the cross term to be in the original form.*

The solution of the LMI in (4.30) or equivalently the values of the variables $\hat{\eta}$ and Π do not have direct influence on the controller design and the actual ultimate bound on the system state and/or sliding function, however, these parameters would lead us to determine a more accurate bound. Therefore, to obtain the minimum value of the bound in (4.29) the LMIs in (4.11) and (4.30) could be solved subject to a specific criteria. This issue is beyond the scope of this book and remains for future works.

4.4 Variable Structure Controller Considerations

As mentioned in [133], $f(k-1)$ is an acceptable estimate of $f(k)$, while the exogenous disturbance is smooth and bounded. Now let us assume that the system signals in (2.1) are transmitted *without* any packet losses, we have

$$f(k-1) = (SB)^{-1}S[x(k) - Ax(k-1) - \Delta A(k-1)x(k-1) - Bu(k-1)].$$

Then, $d(k-1)$ can be estimated by

$$\hat{f}(k) = (SB)^{-1}S[x(k) - Ax(k-1) - Bu(k-1)]. \tag{4.40}$$

The above equation can also be written as

$$\hat{f}(k) = (SB)^{-1}S\Delta A(k-1)x(k-1) + d(k-1).$$

For the balanced uncertainty ΔA, it is known that $(SB)^{-1}S\Delta A(k-1)x(k-1)$ is also balanced and would not change the mean values of the vector $\hat{f}_c(k)$. Besides, while the system state is bounded, the vector $(SB)^{-1}S\Delta A(k-1)x(k-1)$ remains also bounded. Hence, with the proper choice of S and for the small uncertainty ΔA, it can be stated that the magnitude of $(SB)^{-1}S\Delta A(k-1)x(k-1)$ would remain very small compared with that of $f(k-1)$. However, due to the *packet dropout* which exists in the communication network, $\hat{f}(k)$ in (4.40) is not applicable. Hence, the following stochastic disturbance estimator is proposed instead:

$$\hat{f}_c(k) = (SB)^{-1}S[x_c(k) - Ax_c(k-1) - Bu(k-1)]. \tag{4.41}$$

Then, we put the component $\vartheta(k)$ in (4.8) as

$$\vartheta(k) = \mathscr{F}^{+} + \frac{1}{2}\text{diag}(\mathscr{F}^{-})\text{sgn}(\hat{f}_c(k) - \mathscr{F}^{+}), \tag{4.42}$$

where $\text{diag}(\mathscr{F}^{-}) = \text{diag}(f_1^{-}, \cdots, f_m^{-})$. In this case, $\tau = 1.5$ in (2.13) and $\gamma = \frac{9}{4}\|\Pi + (2-\bar{\alpha})SB\|\,\|\mathscr{F}^{-}\|^2$ in (4.29). Note that this bound denotes the worst case scenario bound and, as it is supposed that disturbance in the system (2.1) is slow, this bound, with the perfect position estimation, could be reduced to $\tau^{\star} = 0.5$ and $\gamma^{\star} = \frac{1}{4}\|\Pi + (2-\bar{\alpha})SB\|\,\|\mathscr{F}^{-}\|^2$. Thus, the controller (4.8) can be chosen as

$$u(k) = -(SB)^{-1}SAx_c(k) - \mathscr{F}^{+} - \frac{1}{2}\text{diag}(\mathscr{F}^{-})\text{sgn}(\hat{f}_c(k) - \mathscr{F}^{+}), \tag{4.43}$$

where $\hat{f}_c(k)$ is defined in (4.41). Using the component as in (4.42) in the discrete-time sliding mode controller would lead the state trajectories to chatter around the sliding surfaces with frequency equal to the frequency of the exogenous disturbance. Also, as mentioned, for smooth and bounded exogenous disturbances, it could reduce the thickness of the ultimate bound on the state trajectories and boundary layer around the sliding surface.

4.5 Simulation Results

Again we consider the double integrator system in (1.1), (1.3). Now, using the sampling time of 0.5 s, a discrete-time system is derived as

$$A = \begin{bmatrix} 1.0000 & 0.5000 \\ 0 & 1.0000 \end{bmatrix}, B = \begin{bmatrix} 0.2500 \\ 0.500 \end{bmatrix} \tag{4.44}$$

We also consider the following uncertainty parameters and disturbance in the system

$$\begin{aligned} M &= \begin{bmatrix} 0.0500 & 0.1500 \end{bmatrix}^T, N = \begin{bmatrix} -0.0500 & 0.1000 \end{bmatrix}, \\ R(k) &= 0.5\sin(k). \end{aligned}$$

Suppose

$$f(k) = 0.05\sin\left(\frac{k}{5}\right).$$

The probability of the packet loss is $\bar{\alpha} = 0.1$. Solving the LMI in (4.11) gives the following results:

$$\begin{aligned} P &= \begin{bmatrix} 16.6200 & -1.2487 \\ -1.2487 & 18.4931 \end{bmatrix}, Q = \begin{bmatrix} 4.2899 & 2.2448 \\ 2.2448 & 13.9253 \end{bmatrix} \\ S &= \begin{bmatrix} 3.5307 & 8.9344 \end{bmatrix}, \eta = 0.3928, \varepsilon = 14.8112. \end{aligned}$$

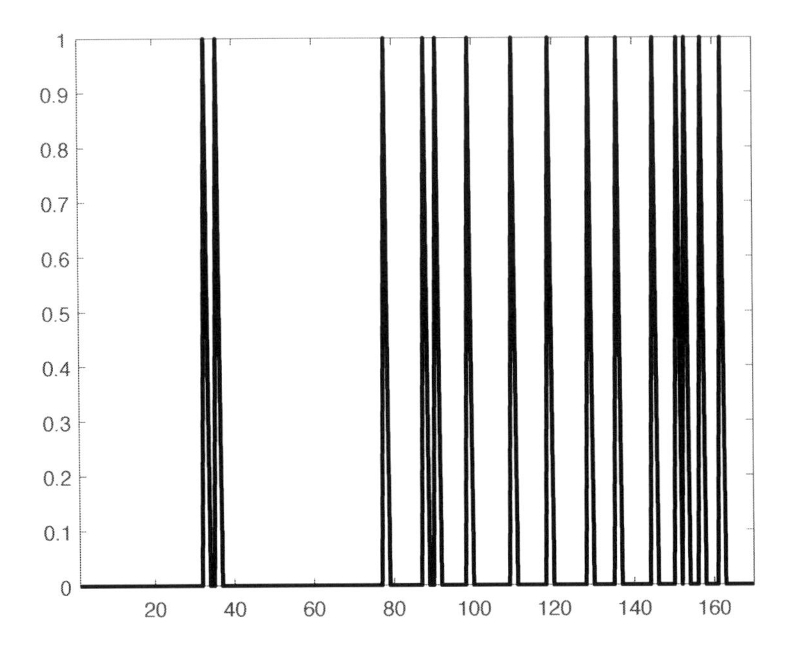

FIGURE 4.2
Bernoulli sequence $\alpha(k)$.

Using S, $\mathscr{F}^+ = 0$ and $\mathscr{F}^- = 0.05$, the control law given in (4.43) can be obtained. Here, the initial state is assumed to be $x(0) = \begin{bmatrix} 2 & -1 \end{bmatrix}^T$. Bernoulli sequence $\alpha(k)$ is depicted in Fig. 4.2.

We apply the controller in (4.43) to the system (2.1). The results are demonstrated in Fig. 4.3.

Now, let us consider the following controller:

$$u(k) = -(SB)^{-1}SAx_c(k) - \mathscr{F}^+ - \text{diag}(\mathscr{F}^-)\text{sgn}(\sigma_x(k)), \tag{4.45}$$

which uses a similar switching component as in [101]. Fig. 4.4 demonstrates the results of using the controller (4.45). Moreover, Fig. 4.5 shows the results of applying only the linear part of the controller (4.43) to the system (2.1).

As seen from these results, the proposed DSMC in (4.43) that uses the practical switching function (4.42), by employing the proposed disturbance estimator (4.41), has effectively better performance in terms of reducing the thickness of the bound on the system state and sliding function in comparison to the one in (4.45) and the linear controller. However, it is worth mentioning that if we do not have any knowledge of the exogenous disturbance in the system, providing the discrete-time sliding mode controller with a switching component cannot necessarily improve the control performance and even can be detrimental to the control performance [67, 127, 98].

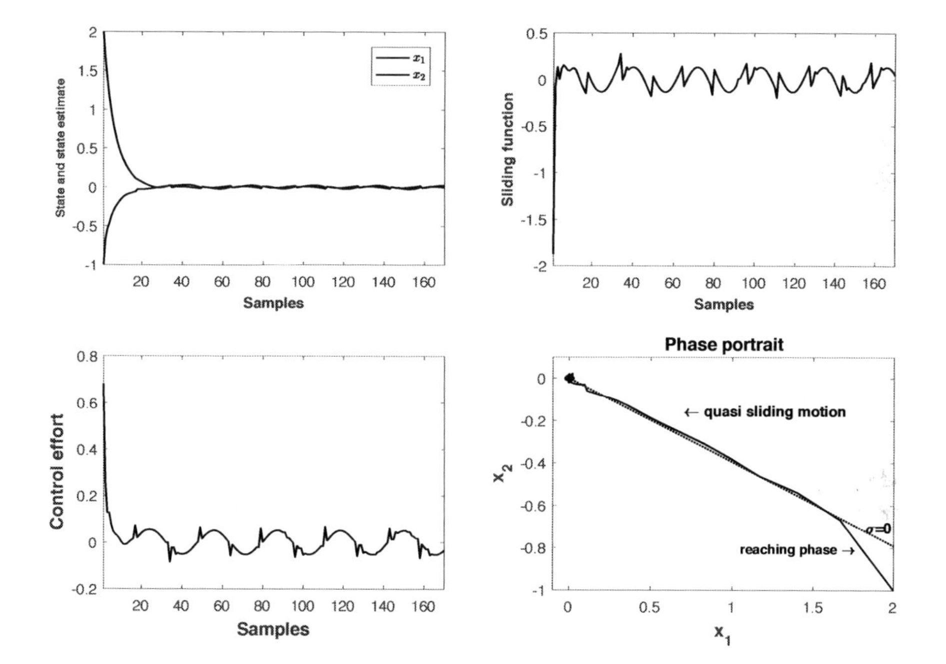

FIGURE 4.3
Results of applying DSMC in (4.43).

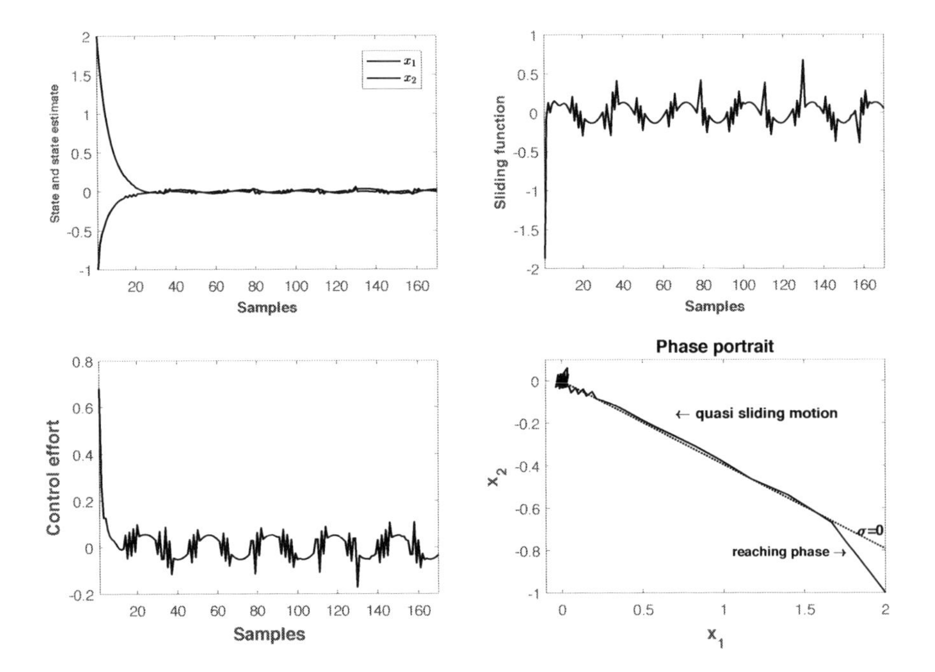

FIGURE 4.4
Results of applying DSMC in (4.45).

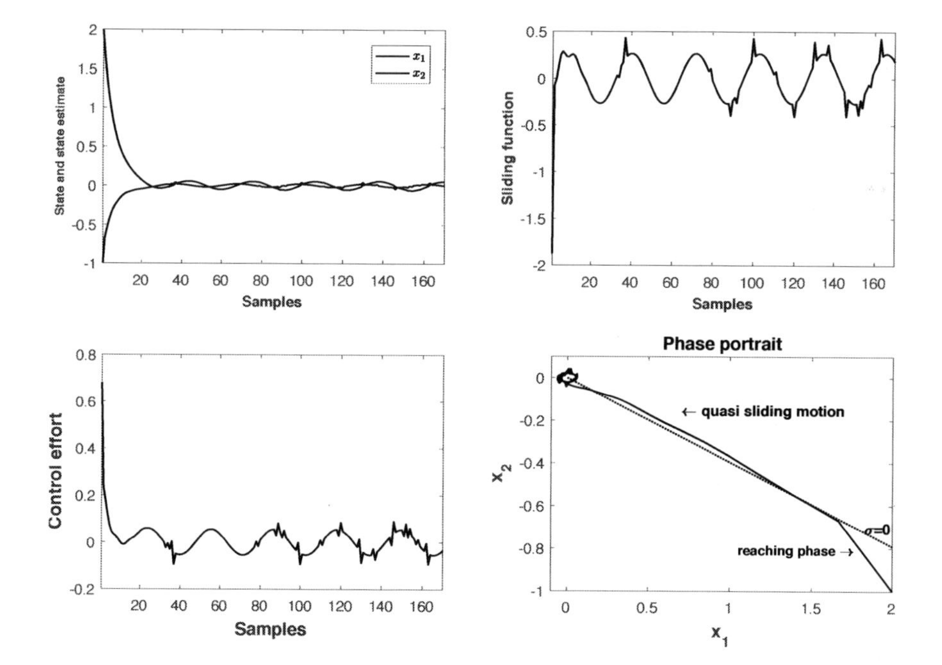

FIGURE 4.5
Results of applying linear controller.

Nevetheless, as seen here, a better performance can be achieved through switching components, when either the sampling rate of the system is very high compared with the maximum frequency component of the exogenous disturbance or the exogenous disturbance is slow (smooth and bounded). With either of these assumptions, the closed-loop system behaves to some extent as a continuous-time system [98] and hence, using a discontinuous component in the controller may improve the performance. Nevertheless, broadly speaking, the large switching gains in the DSMC may result in excessive actions in the control system actuators, hence, it is worth mentioning that the switching gain should be chosen small.

4.6 Conclusions

In this chapter, a stochastic discrete-time sliding mode control has been developed for networked systems involving consecutive measuring data packet losses. Furthermore, using a unified framework, an LMI scheme is developed for the design of a robust sliding surface and controller for the uncertain discrete-time networked systems involving packet dropout. The proposed robust DSMC is applicable to unstable systems, and also there is no need to stabilize the underlying system in advance. Additionally, it could reduce the conservatism of the existing methods in the literature. Using the notion of exponentially mean square stability, the stability and ultimate boundedness of the derived closed-loop system have been analyzed.

5

DSMC for NCSs involving actuation and measurement consecutive packet losses

CONTENTS

5.1 Introduction 87
5.2 Problem Formulation and Preliminaries 88
5.2.1 Problem statement 88
5.3 Stochastic Sliding Mode Control 91
5.3.1 Designing the sliding function subject to consecutive packet losses 91
5.3.2 Stability analysis 93
5.4 Numerical Examples 103
5.4.1 Example 1 103
5.5 Conclusions 104

Abstract–This chapter considers the problem of designing a robust output feedback discrete-time sliding mode control (ODSMC) for the networked systems involving both measuring and actuating data packet losses. Packet losses in the networked control systems (NCSs) have been modeled by utilizing the probability and the characteristics of the sources and the destinations. Here, the Bernoulli random binary distribution is used to model consecutive packet losses in the NCSs. In this chapter, first, a robust observer-based discrete-time sliding mode control is proposed for the NCSs including random packet losses. The packet losses occur in the channels from the sensors to the controller and the channels from the controller to the actuators. Then, using the notion of exponential mean square stability, the boundedness of the obtained closed-loop system is analyzed with an LMI approach. Illustrative examples are presented to show the effectiveness of the proposed approach.

5.1 Introduction

The DSMC given in the previous chapter is derived based on two assumptions:

1. the packet losses occur only in the channel from the sensor to the controller;
2. the system states are entirely available.

Sometimes, these are not clearly very realistic assumptions for many of practical problems. This chapter intends to design sliding mode controllers for the NCSs involving both measurement and actuation consecutive packet losses (or long-term random delays), which exploit only output information. This ODSMC can distinguish itself from the existing literature on the SMCs applied to the NCSs, in the sense that both the measurement and actuation delays are viewed as the Bernoulli distributed white sequence. This matter has been considered for the design of other control strategies such as observer-based H_∞ and state feedback controllers [146, 91].

In brief, the main goal of this chapter is to stabilize a NCS involving measurement and actuation packet dropouts with an observer-based ODSMC. The main contribution of this chapter includes the following major innovations.

- This chapter revises the observer utilized in [146, 91] to a more practical alternative. Indeed, [146, 91] assume that the state observer and controller are not located in the same place, or equivalently, the control signals used in the state observer involve the random time-delays that exist in the channel from the system controller to the actuators. This could not be a real assumption in most of the NCSs.
- The proposed ODSMC can be applied to NCSs involving both the measuring and actuating consecutive packet losses.
- The proposed robust ODSMC scheme provides an integrated framework for general systems. This is certainly different from the DSMC introduced in [101] which can only be applied to the stable systems, and [102] which stabilizes the underlying system first and then designs the sliding mode controller separately.
- A novel method is developed to reduce the conservatism which exists in the current literature that removes the cross term between the system state and disturbance to make a fully diagonal problem.

The rest of this chapter is organized as follows: Section 5.2 describes the problem formulation. Section 5.3 presents the proposed method to design the sliding surface and ODSMC. Effectiveness of the proposed ODSMC is studied by numerical examples in Section 5.4. Finally, Section 5.5 concludes this chapter.

5.2 Problem Formulation and Preliminaries

5.2.1 Problem statement

Consider the NCS with (consecutive) random packet losses shown in Fig. 5.1. As seen the NCS here involves both measurement and actuation packet dropout. In

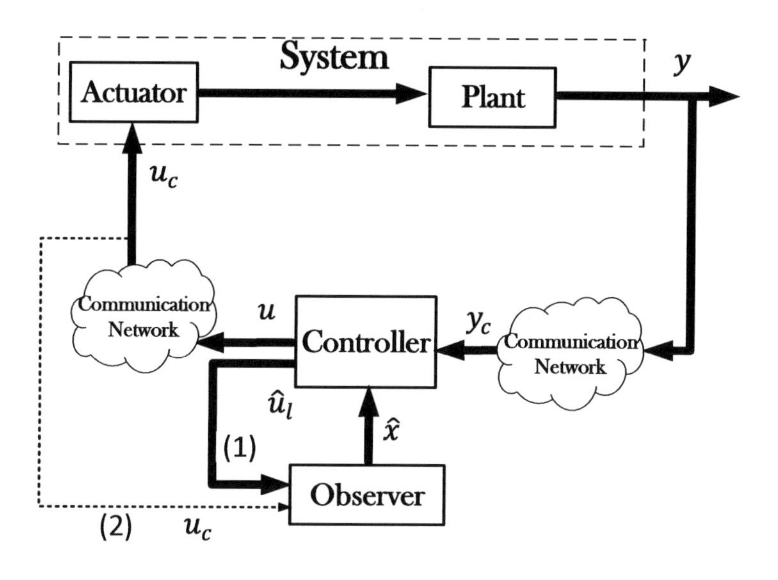

FIGURE 5.1
NCS structure

other words, the system does not have access to the control input $u(k)$, generated by the controller, and instead it would be utilized with the communicated control input $u_c(k)$. On the other hand, due to the packet losses existing in the channel from the sensors to the controller, the observer, which is at the same location of the controller, also does not receive the actual values of the system output $y(k)$, but $y_c(k)$. The observer will provide the state estimate $\hat{x}(k)$ for the controller using the signal $\hat{u}_l(k)$, which will be defined later in this chapter, rather than $u(k)$ (*cf.* [146, 91]).

Now, consider the following uncertain linear discrete-time system which explains the dynamics in Fig. 5.1,

$$\textbf{System}: \begin{cases} x(k+1) = [A + \Delta A(k)]x(k) + B[u_c(k) + f(k)] \\ y(k) = Cx(k), \end{cases} \tag{5.1}$$

where $x(k) \in \mathbb{R}^n$, $u_c(k) \in \mathbb{R}^m$ and $y(k) \in \mathbb{R}^p$ are system state, communicated control input (see Fig. 5.1) and system output respectively. Without loss of generality, it is assumed that $m \leq n$, $\text{rank}(B) = m$ and $\text{rank}(C) = p$. Besides, it is assumed that (A, B) is controllable and (A, C) is observable. The uncertain matrix $\Delta A(k)$ has the form of (2.2).

As the measured outputs are sent to the controller via a communication network, the measurements may involve randomly varying communication delays and/or the detrimental phenomenon referred to as *data packet dropout*. The communicated output $y_c(k)$ and communicated input $u_c(k)$ are assumed to be as

$$\textbf{Communicated output}: \begin{cases} x_c(k) = (1-\alpha(k))x(k) + \alpha(k)x_c(k-1) \\ y_c(k) = Cx_c(k), \end{cases} \tag{5.2}$$

$$\textbf{Communicated input}: \ u_c(k) = (1-\beta(k))u(k), \tag{5.3}$$

where $x_c(k)$ denotes the communicated system state which is not available and the stochastic variables $\alpha(k) \in \mathbb{R}$ and $\beta(k) \in \mathbb{R}$ are Bernoulli distributed white sequences with

$$\begin{cases} \text{Prob}\{\alpha(k)=1\} = \mathbb{E}\{\alpha(k)\} = \bar{\alpha} \\ \text{Prob}\{\alpha(k)=0\} = 1-\mathbb{E}\{\alpha(k)\} = 1-\bar{\alpha}, \end{cases} \tag{5.4}$$

$$\begin{cases} \text{Prob}\{\beta(k)=1\} = \mathbb{E}\{\beta(k)\} = \bar{\beta} \\ \text{Prob}\{\beta(k)=0\} = 1-\mathbb{E}\{\beta(k)\} = 1-\bar{\beta}, \end{cases} \tag{5.5}$$

in which $0 \le \bar{\alpha} < 1$, $0 \le \bar{\beta} < 1$ imply the probability that a data packet may be lost or have delay. The stochastic variable $\beta(k)$ is mutually independent of the stochastic variable $\alpha(k)$. Notice that $u_c(k)$ involves a random communication delay similar to that in the measurement communication, however, with a different switching probability. It can be frequently seen in the literature to use a similar Bernoulli distributed white sequence model but with different switching probability for the random delays in the signals from the sensors to the controller and also in the signals from the controller to the actuators; see e.g. [146, 91]. Besides, two different schemes for the systems involving packet loss have been considered, i.e., the hold-input method (which is used here in the communicated output $y_c(k)$ in (5.2)) and the zero-input method (which is used here in the communicated input $u_c(k)$ in (5.3)) [119, 126, 118]. The hold-input method seems to have better performance than the zero-input one. However, according to the existing literature (e.g. see [119, 118]), either of them cannot outperform the other, even in simple scalar systems [118]. Therefore, here, for simplification purposes, the zero-input method would be utilized to deal with the packet dropouts in the channel from the controller to the actuators.

In this chapter, we use the following estimation scheme to provide the controller with the system state information,

$$\textbf{Observer}: \begin{cases} \hat{x}(k+1) = A\hat{x}(k) + B\hat{u}_l(k) + \frac{1}{1-\bar{\alpha}}L[y_c(k) - \hat{y}_c(k)] \\ \hat{u}_l(k) = (1-\bar{\beta})u_l(k) \\ \hat{y}_c(k) = (1-\bar{\alpha})C\hat{x}(k) + \bar{\alpha}C\hat{x}(k-1), \end{cases} \tag{5.6}$$

where $\hat{x}(k) \in \mathbb{R}^n$ is the state estimate of the system in (5.1), $\hat{y}_c(k) \in \mathbb{R}^p$ is the observer output, $L \in \mathbb{R}^{n \times p}$ is the observer gain and $u_l(k) \in \mathbb{R}^m$ denotes the linear part of the system controller $u(k)$ proposed in the rest of this chapter.

Remark 5.1 *The measurement model* (5.2) *is fundamentally different from the one that has been used in [113, 141, 146, 91] etc. Note that in the literature, rather than the measurement model* (5.2)*, the following one is frequently utilized,*

$$y_c(k) = (1-\alpha(k))y(k) + \alpha(k)y(k-1). \tag{5.7}$$

This model indeed would impose an assumption on the system networks that the packet dropout may not occur successively. This is obviously not a realistic assumption. Instead, the model (5.2) *can cope with the longer random delays and/or frequent packet losses. Notice that the model* (5.7) *has been used to cope with both time varying communication delay (e.g. see [146, 91]) and data packet dropout (e.g. see [101]). Considering the model* (5.2) *instead of model* (5.7)*, if a packet loss occurs, the controller is provided with* $y_c(k) = y_c(k-1)$*, or else the controller utilizes* $y_c(k) = y(k)$ *instead. Here,* $y_c(k-1)$ *denotes the last available communicated data packet. This justification can also be used for the communication random time-delay accordingly. It means that if the time-delay of the communication channel, which is assumed to be* τ_d*, is less than a sampling period* (T) *of the discrete-time control system, the delay has no influence on the system and we have* $y_c(k) = y(k)$*. But, if* $\tau_d \geq T$*, then* $y_c(k) = y_c(k-1)$*. Note that, in the case of random delay occurrence,* $y_c(k-1)$ *can be regarded as* $y(k-\tau_d)$*.*
Notice also that we do not bound the number of possible consecutive packet losses (or equivalently the random time-delays cf. [91]). Indeed, there is an implicit stochastic constraint indicated by the Bernoulli variable $\alpha(k)$ *as* $Prob\{\alpha(k) = 1\} = \mathbb{E}\{\alpha(k)\} = \bar{\alpha}$*. In simpler terms, if* $\bar{\alpha}$ *is a large value, the number of possible consecutive packet losses increases and vice versa.*

Remark 5.2 *Notice that in [146, 91], the observer in* (5.6) *is assumed to be as*

$$\hat{x}(k+1) = A\hat{x}(k) + Bu_{l_c}(k) + L[y_c(k) - \hat{y}_c(k)],$$

where u_{l_c} *denotes the linear part of* $u_c(k)$*. In other words, the control input utilized in the observer is the (linear part of the) communicated control input to the system which involves random communication delay. This itself means that either the observer and controller are not located in the same place together, or the observer has access to* $u_c(k)$*; see the dashed line in Fig 5.1. This structure does not seem to be a practical case in most of the NCSs. Therefore, to address this problem, in this chapter, the control input* $\hat{u}_l(k)$ *is exploited in the observer (see* (5.6)*), which is more applicable for practical NCSs.*

This chapter will consider the problem of designing a robust observer-based DSMC for the NCS in (5.1)-(5.3). In the rest of this chapter, for simplification, the brief α and ΔA are used instead of $\alpha(k)$ and $\Delta A(k)$, respectively.

5.3 Stochastic Sliding Mode Control

This section aims to design a robust observer-based ODSMC in order to stabilize the NCS in (5.1).

5.3.1 Designing the sliding function subject to consecutive packet losses

Consider the following discrete-time linear sliding function,

$$\sigma(k) = S\hat{x}(k), \tag{5.8}$$

where the matrix $S \in \mathbb{R}^{m\times n}$ will be designed later such that SB is nonsingular. Note that in the ideal sliding mode we have

$$\sigma(k) = 0, \ \ \forall k \geq k_s, \tag{5.9}$$

where $k_s > 0$ denotes the time when sliding motion starts. The controller is assumed to be of the following structure,

$$u(k) = -\frac{1}{1-\bar{\beta}}[(SB)^{-1}(SA - \Phi S)\hat{x}(k) + \vartheta(k)], \tag{5.10}$$

where $\Phi \in \mathbb{R}^{m\times m}$ is a known stable matrix and $\vartheta(k)$ denotes the approximation of the disturbance $f(k)$ used in the controller to compensate the harmful effect of disturbance on the ultimate bound on the system state trajectories. In addition to the assumption in (2.13), we also assume

$$\|\vartheta(k)\| \leq \kappa \left\| \mathscr{F}^b \right\|, \tag{5.11}$$

where $\kappa > 0$ is a scalar. The term $(SB)^{-1}\Phi S\hat{x}(k)$, in $u(k)$ (see (5.10)), is used to govern the rate of convergence to the sliding manifold. Here, similar to [40], it is assumed that $\Phi = \lambda I_m$, where $0 \leq \lambda < 1$ is a given constant value. Thanks to the special form of Φ which can commute with S, the control law $u(k)$ in (5.10) could be written as

$$\begin{aligned}\textbf{Controller}: \ u(k) &= -\frac{1}{1-\bar{\beta}}[(SB)^{-1}SA_\lambda\hat{x}(k) + \vartheta(k)] \\ &\triangleq u_l(k) + u_n(k),\end{aligned} \tag{5.12}$$

where $A_\lambda = A - \lambda I_n$ and $u_l(k)$ is the linear part of $u(k)$ which was used in the observer design (5.6). Define the estimation errors as

$$\textbf{Estimation error}: \begin{cases} e(k) := x(k) - \hat{x}(k) \\ e_c(k) := x_c(k) - \hat{x}(k). \end{cases} \tag{5.13}$$

Then, by applying the controller (5.12) to (5.1) and also using (5.13), (5.2) and (5.6), the closed-loop system is obtained as

Closed-loop system :

$$\begin{cases} x(k+1) = \left[A + \Delta A - \hat{A}_\lambda + \frac{\beta-\bar{\beta}}{1-\bar{\beta}}\hat{A}_\lambda\right]x(k) + \left[1 - \frac{\beta-\bar{\beta}}{1-\bar{\beta}}\right]\hat{A}_\lambda e(k) \\ \qquad\qquad +B\left[d_\vartheta(k) + \frac{\beta-\bar{\beta}}{1-\bar{\beta}}\vartheta(k)\right] \\ e(k+1) = \left[\Delta A + \frac{\beta-\bar{\beta}}{1-\bar{\beta}}\hat{A}_\lambda + \frac{\alpha-\bar{\alpha}}{1-\bar{\alpha}}LC\right]x(k) + \left[A - LC - \frac{\beta-\bar{\beta}}{1-\bar{\beta}}\hat{A}_\lambda\right]e(k) \\ \qquad\qquad -\frac{\alpha-\bar{\alpha}}{1-\bar{\alpha}}LCx_c(k-1) - \pi LCe_c(k-1) + B\left[d_\vartheta(k) + \frac{\beta-\bar{\beta}}{1-\bar{\beta}}\vartheta(k)\right] \\ x_c(k) = (1-\alpha)x(k) + \alpha x_c(k-1) \\ e_c(k) = -\alpha x(k) + \alpha x_c(k-1) + e(k) \end{cases}$$

where $\hat{A}_\lambda = B(SB)^{-1}SA_\lambda$ and $\pi = \frac{\bar{\alpha}}{1-\bar{\alpha}}$. Also, it can simply be found that

$$\begin{aligned}\sigma(k+1) =& S\left(\lambda I - \frac{\alpha - \bar{\alpha}}{1-\bar{\alpha}}LC\right)x(k) - S(\lambda I - LC)e(k) \\ &+ \frac{\alpha - \bar{\alpha}}{1-\bar{\alpha}}SLCx_c(k-1) + \pi SLCe_c(k-1). \end{aligned} \quad (5.14)$$

5.3.2 Stability analysis

Notice that in the case of applying DSMC to discrete-time systems involving exogenous disturbances, the closed-loop system should be analyzed in terms of boundedness. Also, the proposed DSMC could only ensure the state trajectories to be driven into a boundary layer around the ideal sliding surface $\sigma(k) = 0$. This issue is indeed regarded as the quasi sliding mode (QSM) in the literature. The following theorem considers a method to analyze simultaneously the reachabiltiy of QSM and the stability of the system states utilizing a discrete-time Lyapunov stability method, in the absence of exogenous disturbances. The characterization of the bounds on the closed-loop system states and sliding function's boundary layer are presented separately later in Theorem 5.2. Further, as Theorem 5.2 needs to derive the cross terms between the system state (sliding function) and the components $f_\vartheta(k)$ and $\vartheta(k)$, in order to avoid unnecessary repetition of the technical manipulations, we will start the proof of Theorem 5.1 more generally (with the external disturbance and the discontinuous component $\vartheta(k)$) for the sake of Theorem 5.2. We then let $\begin{bmatrix} f_\vartheta(k) \\ \vartheta(k) \end{bmatrix} = 0$ to derive the LMI condition for the stability analysis and control and observer synthesis.

Theorem 5.1 *In the absence of the exogenous disturbance $f(k)$, the control law* (5.10) *can steer the state of the stochastic system* (5.1) *onto the ideal sliding surface* (5.9) *and, also the system state is exponentially mean square stable if there exist matrices $0 < P := U^T \begin{bmatrix} P_{11} & 0 \\ 0 & P_{22} \end{bmatrix} U$, $Q_1 > 0$, $Q_2 > 0$, X_1, X_2, X_3 and X_4, and scalars*

$\varepsilon > 0$ and $\rho > 0$ satisfying the following LMI:

$$\begin{bmatrix}
\check{\Sigma}_{11} & \star & \star & \star & \star & \star & \star & \star & \star & \star & \star & \star & \star \\
\tilde{\Sigma}_{12}^T & \tilde{\Sigma}_{22} & \star & \star & \star & \star & \star & \star & \star & \star & \star & \star & \star \\
BX_2 & BX_3 & -P & \star & \star & \star & \star & \star & \star & \star & \star & \star & \star \\
-\bar{\alpha}Q_2 & \bar{\alpha}Q_2 & 0 & \bar{\Sigma}_{33} & \star & \star & \star & \star & \star & \star & \star & \star & \star \\
0 & 0 & 0 & 0 & -Q_2+\rho I & \star & \star & \star & \star & \star & \star & \star & \star \\
\lambda B^T P & B^T PA_\lambda & 0 & 0 & 0 & -B^T PB & \star & \star & \star & \star & \star & \star & \star \\
-\lambda B^T P & B^T(\lambda P - X_4C) & 0 & 0 & -\pi B^T X_4 C & 0 & -B^T PB & \star & \star & \star & \star & \star & \star \\
-2\phi B^T PA_\lambda & 2\phi B^T PA_\lambda & 0 & 0 & 0 & 0 & 0 & -2\phi B^T PB & \star & \star & \star & \star & \star \\
0 & PA - X_4C & 0 & 0 & -\pi X_4C & 0 & 0 & 0 & -P & \star & \star & \star & \star \\
\pi X_4C & 0 & 0 & -\pi X_4C & 0 & 0 & 0 & 0 & 0 & -\pi P & \star & \star & \star \\
\pi B^T X_4C & 0 & 0 & -\pi B^T X_4C & 0 & 0 & 0 & 0 & 0 & 0 & -\pi B^T PB & \star & \star \\
PA+BX_1 & 0 & 0 & 0 & 0 & 0 & 0 & 0 & 0 & 0 & 0 & -P & \star \\
0 & 0 & 0 & 0 & 0 & M^T PB & 0 & 0 & M^T P & 0 & 0 & M^T P & -\varepsilon I
\end{bmatrix} < 0, \quad (5.15)$$

where M and N are known matrices of the uncertainty in (2.2), $0 < P_{22} \in \mathbb{R}^{(n-m)\times(n-m)}$ *and $U \in \mathbb{R}^{n\times n}$ is defined in Lemma 3.1,* $\check{\Sigma}_{11} = -P + (1-\bar{\alpha})Q_1 + \bar{\alpha}Q_2 + X_2^T B^T + BX_2 + \rho I + \varepsilon N^T N$, $\tilde{\Sigma}_{12} = -X_2^T B^T + BX_3 - \bar{\alpha}Q_2$, $\tilde{\Sigma}_{22} = -X_3^T B^T - BX_3 - P + Q_2 + \rho I$, $\bar{\Sigma}_{33} = -(1-\bar{\alpha})Q_1 + \bar{\alpha}Q_2 + \rho I$, $\pi = \frac{\bar{\alpha}}{1-\bar{\alpha}}$, $\phi = \frac{\bar{\beta}}{1-\bar{\beta}}$ *and $\{\star\}$ denotes the symmetric elements in a symmetric matrix. Here $S = B^T P$ and the observer gain is*

$$L = P^{-1}X_4. \quad (5.16)$$

Proof 13 *Define*

$$\begin{aligned}
V(\zeta(k)) =& x^T(k)Px(k) + e^T(k)Pe(k) + x_c^T(k-1)Q_1x_c(k-1) \\
&+ e_c^T(k-1)Q_2e_c(k-1) + \sigma^T(k)(SB)^{-1}\sigma(k),
\end{aligned} \quad (5.17)$$

where $\zeta(k) = \left[x^T(k)\ e^T(k)\ x_c^T(k-1)\ e_c^T(k-1)\ \sigma^T(k)\right]^T$, $P > 0$, $Q_1 > 0$ *and* $Q_2 > 0$ *are sym-*

metric matrices and $S = B^T P$*. Thus, it can be written*

$$\begin{aligned}
&\Delta V(\zeta(k)) \triangleq \mathbb{E}\{V(\zeta(k+1))|\zeta(k)\} - V(\zeta(k)) \\
=&\mathbb{E}\{x^T(k+1)Px(k+1) + e^T(k+1)Pe(k+1) + x_c^T(k)Q_1x_c(k) + e_c^T(k)Q_2e_c(k) \\
&+ \sigma^T(k+1)(SB)^{-1}\sigma(k+1)|\zeta(k)\} - x^T(k)Px(k) - e^T(k)Pe(k) \\
&- x_c^T(k-1)Q_1x_c(k-1) - e_c^T(k-1)Q_2e_c(k-1) - \sigma^T(k)(SB)^{-1}\sigma(k).
\end{aligned} \tag{5.18}$$

It is then followed by

$$\begin{aligned}
&\mathbb{E}\{x^T(k+1)Px(k+1)|\zeta(k)\} = \mathbb{E}\left\{x^T(k)\left[A + \Delta A - \hat{A}_\Delta + \Delta A_\lambda + \frac{\beta - \bar{\beta}}{1 - \bar{\beta}}\hat{A}_\lambda\right]^T\right. \\
&\left.\times P\left[A + \Delta A - \hat{A}_\Delta + \Delta A_\lambda + \frac{\beta - \bar{\beta}}{1 - \bar{\beta}}\hat{A}_\lambda\right]x(k)|\zeta(k)\right\} \\
&+ 2x^T(k)[(\lambda I_n + \Delta A)^T S^T(SB)^{-1}SA_\lambda - \phi A_\lambda^T S^T(SB)^{-1}SA_\lambda]e(k) \\
&+ 2x^T(k)(\Delta A + \lambda I_n)^T S^T f_\vartheta(k) \\
&+ 2\phi x^T(k)A_\lambda^T S^T\vartheta(k) + 2e^T(k)A_\lambda^T S^T f_\vartheta(k) - 2\phi e^T(k)A_\lambda^T S^T\vartheta(k) \\
&+ (1+\phi)e^T(k)A_\lambda^T S^T(SB)^{-1}SA_\lambda e(k) + f_\vartheta(k)^T(SB)f_\vartheta(k) + \phi\vartheta^T(k)(SB)\vartheta(k) \\
=&x^T(k)[(A+\Delta A)^T P(A+\Delta A) - (A+\Delta A)^T S^T(SB)^{-1}S(A+\Delta A) \\
&+ (\lambda I_n + \Delta A)^T S^T(SB)^{-1}S(\lambda I_n + \Delta A + \phi A_\lambda^T S^T(SB)^{-1}SA_\lambda]x(k)) \\
&+ 2x^T(k)[(\lambda I_n + \Delta A)^T S^T(SB)^{-1}SA_\lambda - \phi A_\lambda^T S^T(SB)^{-1}SA_\lambda]e(k) \\
&+ 2x^T(k)(\Delta A + \lambda I_n)^T S^T f_\vartheta(k) + 2\phi x^T(k)A_\lambda^T S^T\vartheta(k) + 2e^T(k)A_\lambda^T S^T f_\vartheta(k) \\
&- 2\phi e^T(k)A_\lambda^T S^T\vartheta(k) + (1+\phi)e^T(k)A_\lambda^T S^T(SB)^{-1}SA_\lambda e(k) + f_\vartheta(k)^T(SB)f_\vartheta(k) \\
&+ \phi\vartheta^T(k)(SB)\vartheta(k)
\end{aligned} \tag{5.19}$$

where $\hat{A}_\Delta = B(SB)^{-1}S(A+\Delta A)$*,* $\Delta A_\lambda = B(SB)^{-1}S(\lambda I_n + \Delta A)$ *and* $\phi = \frac{\bar{\beta}}{1-\bar{\beta}}$*. Besides, note that* $\mathbb{E}\{(\beta - \bar{\beta})\} = 0$ *and* $\mathbb{E}\{(\beta - \bar{\beta})^2\} = \bar{\beta}(1-\bar{\beta}) \triangleq \delta^2$*. Thus, it can be demonstrated that*

$$\begin{aligned}
&\mathbb{E}\{e^T(k+1)Pe(k+1)|\zeta(k)\} \\
=&x^T(k)[\Delta A^T P\Delta A + \phi A_\lambda^T PB(B^T PB)^{-1}B^T PA_\lambda + \pi(LC)^T PLC]x(k) \\
&+ e^T(k)[(A-LC)^T P(A-LC) + \phi A_\lambda^T PB(B^T PB)^{-1}B^T PA_\lambda]e(k) \\
&+ \pi x_c^T(k-1)(LC)^T PLCx_c(k-1) + \pi^2 e_c^T(k-1)(LC)^T PLCe_c(k-1) \\
&+ f_\vartheta^T(k)B^T PBf_\vartheta(k) + \phi\vartheta^T(k)B^T PB\vartheta(k) \\
&+ 2x^T(k)[\Delta A^T P(A-LC) - \phi A_\lambda^T PB(B^T PB)^{-1}B^T PA_\lambda]e(k) \\
&- 2\pi x^T(k)(LC)^T PLCx_c(k-1) - 2\pi x^T(k)\Delta A^T PLCe_c(k-1) + 2x^T(k)\Delta A^T PBf_\vartheta(k) \\
&+ 2\phi x^T(k)A_\lambda^T PB\vartheta(k) - 2\pi e^T(k)(A-LC)^T PLCe_c(k-1) \\
&+ 2e^T(k)(A-LC)^T PBf_\vartheta(k) - 2\phi e^T(k)A_\lambda^T PB\vartheta(k) - 2\pi e_c^T(k-1)C^T L^T PBf_\vartheta(k),
\end{aligned} \tag{5.20}$$

in which again $\mathbb{E}\{(\alpha-\bar{\alpha})\}=0$ *and* $\mathbb{E}\{(\alpha-\bar{\alpha})^2\}=\bar{\alpha}(1-\bar{\alpha})\triangleq\psi^2$. *Also*

$$\begin{aligned}&\mathbb{E}\left\{x_c^T(k)Q_1x_c(k)\middle|\zeta(k)\right\}\\=&(1-\bar{\alpha})x^T(k)Q_1x(k)+\bar{\alpha}x_c^T(k-1)Q_1x_c(k-1),\end{aligned}\tag{5.21}$$

and

$$\begin{aligned}&\mathbb{E}\{e_c^T(k)Q_2e_c(k)\big|\zeta(k)\}\\=&\mathbb{E}\{[-\alpha x(k)+\alpha x_c(k-1)+e(k)]^TQ_2[-\alpha x(k)+\alpha x_c(k-1)+e(k)]\big|\zeta(k)\}\\=&\bar{\alpha}x^T(k)Q_2x(k)-2\bar{\alpha}x^T(k)Q_2x_c(k-1)-2\bar{\alpha}x^T(k)Q_2e(k)+\bar{\alpha}x_c^T(k-1)Q_2x_c(k-1)\\&+2\bar{\alpha}x_c^T(k-1)Q_2e(k)+e^T(k)Q_2e(k),\end{aligned}\tag{5.22}$$

in which $\mathbb{E}\{\alpha^2\}=\bar{\alpha}$. *Besides, we have*

$$\begin{aligned}&\mathbb{E}\{\sigma^T(k+1)(SB)^{-1}\sigma(k+1)\big|\zeta(k)\}\\=&x^T(k)[\pi C^TL^TS^T(SB)^{-1}SLC+\lambda^2S^T(SB)^{-1}S]x(k)\\&+e^T(k)(\lambda I-LC)^TS^T(SB)^{-1}S(\lambda I-LC)e(k)\\&+\pi x_c^T(k-1)C^TL^TS^T(SB)^{-1}SLCx_c(k-1)\\&+\pi^2e_c^T(k-1)C^TL^TS^T(SB)^{-1}SLCe_c(k-1)\\&-2x^T(k)\lambda S^T(SB)^{-1}S(\lambda I-LC)e(k)-2\pi x^T(k)C^TL^TS^T(SB)^{-1}SLCx_c(k-1)\\&+2\pi\lambda x^T(k)S^T(SB)^{-1}SLCe_c(k-1)\\&-2\pi e^T(k)(\lambda I-LC)^TS^T(SB)^{-1}SLCe_c(k-1).\end{aligned}\tag{5.23}$$

It is also easy to show that

$$\begin{aligned}&\sigma^T(k)(SB)^{-1}\sigma(k)=[x(k)-e(k)]^TS^T(SB)^{-1}S[x(k)-e(k)]\\&=x^T(k)S^T(SB)^{-1}Sx(k)-2x^T(k)S^T(SB)^{-1}Se(k)+e^T(k)S^T(SB)^{-1}Se(k).\end{aligned}\tag{5.24}$$

Using (5.19)-(5.24) *and defining*

$$\varpi(k)=\begin{bmatrix}x^T(k) & e^T(k) & x_c^T(k-1) & e_c^T(k-1) & f_\vartheta^T(k) & \vartheta^T(k)\end{bmatrix}^T,$$

Equation (5.18) *can be rearranged as*

$$\Delta V(\zeta(k))=\varpi^T(k)\,[\Sigma_{ij}]_{6\times6}\,\varpi(k),\tag{5.25}$$

where

$$\begin{aligned}
\Sigma_{11} =&(A+\Delta A)^T P(A+\Delta A) - (A+\Delta A)^T S^T (SB)^{-1} S(A+\Delta A) \\
&+ (\lambda I_n + \Delta A)^T S^T (SB)^{-1} S(\lambda I_n + \Delta A) \\
&+ \Delta A^T P \Delta A + 2\phi A_\lambda^T PB(B^T PB)^{-1} B^T PA_\lambda - (1-\lambda^2) S^T (SB)^{-1} S + \\
&\pi (LC)^T PLC + \pi C^T L^T S^T (SB)^{-1} SLC - P + (1-\bar{\alpha}) Q_1 + \bar{\alpha} Q_2, \\
\Sigma_{12} =&(\lambda I_n + \Delta A)^T S^T (SB)^{-1} SA_\lambda + \Delta A^T P(A-LC) + S^T (SB)^{-1} S \\
&- 2\phi A_\lambda^T PB(B^T PB)^{-1} B^T PA_\lambda - \bar{\alpha} Q_2 - \lambda S^T (SB)^{-1} S(\lambda I - LC), \\
\Sigma_{13} =& -\pi (LC)^T PLC - \pi C^T L^T S^T (SB)^{-1} SLC - \bar{\alpha} Q_2, \\
\Sigma_{14} =& -\pi \Delta A^T PLC + \pi \lambda S^T (SB)^{-1} SLC, \\
\Sigma_{15} =&(\Delta A + \lambda I_n)^T S^T + \Delta A^T PB, \\
\Sigma_{16} =&2\phi A_\lambda^T PB, \\
\Sigma_{22} =&(1+2\phi) A_\lambda^T S^T (SB)^{-1} SA_\lambda + (\lambda I - LC)^T S^T (SB)^{-1} S(\lambda I - LC) \\
&+ (A-LC)^T P(A-LC) - S^T (SB)^{-1} S - P + Q_2, \\
\Sigma_{23} =&\bar{\alpha} Q_2, \\
\Sigma_{24} =& -\pi (A-LC)^T PLC - \pi (\lambda I - LC)^T S^T (SB)^{-1} SLC, \\
\Sigma_{25} =&A_\lambda^T S^T + (A-LC)^T PB, \\
\Sigma_{26} =& -2\phi A_\lambda^T PB, \\
\Sigma_{33} =&\pi (LC)^T PLC + \pi C^T L^T S^T (SB)^{-1} SLC + \bar{\alpha} Q_2 - (1-\bar{\alpha}) Q_1, \\
\Sigma_{34} =&0, \\
\Sigma_{35} =&0, \\
\Sigma_{36} =&0, \\
\Sigma_{44} =&\pi^2 (LC)^T PLC + \pi^2 C^T L^T S^T (SB)^{-1} SLC - Q_2,
\end{aligned}$$

and $\Sigma_{45} = -\pi C^T L^T PB$, $\Sigma_{46} = 0$, $\Sigma_{55} = 2SB$, $\Sigma_{56} = 0$, $\Sigma_{66} = 2\phi SB$. *Now, to prove the system stability, let* $\begin{bmatrix} f_\vartheta(k) \\ \vartheta(k) \end{bmatrix} = 0$. *Then the system is stable if*

$$\Xi := [\Sigma_{ij}]_{4\times 4} < -\rho I, \tag{5.26}$$

where $\rho > 0$ *is a scalar variable. With the choice of* $S = B^T P$ *and utilizing the Schur complement, it can be shown that the feasibility of* (5.26) *is equivalent to that of*

$$\begin{bmatrix} \bar{\Sigma}_{11} & \star & \star & \star & \star & \star & \star & \star & \star & \star \\ \bar{\Sigma}_{12}^T & \bar{\Sigma}_{22} & \star & \star & \star & \star & \star & \star & \star & \star \\ -\bar{\alpha}Q_2 & \bar{\alpha}Q_2 & \bar{\Sigma}_{33} & \star & \star & \star & \star & \star & \star & \star \\ 0 & 0 & 0 & -Q_2+\rho I & \star & \star & \star & \star & \star & \star \\ B^T P(\lambda I_n+\Delta A) & B^T PA_\lambda & 0 & 0 & -B^T PB & \star & \star & \star & \star & \star \\ -\lambda B^T P & B^T P(\lambda I-LC) & 0 & -\pi B^T PLC & 0 & -B^T PB & \star & \star & \star & \star \\ -2\phi B^T PA_\lambda & 2\phi B^T PA_\lambda & 0 & 0 & 0 & 0 & -2\phi B^T PB & \star & \star & \star \\ P\Delta A & P(A-LC) & 0 & -\pi PLC & 0 & 0 & 0 & -P & \star & \star \\ \pi PLC & 0 & -\pi PLC & 0 & 0 & 0 & 0 & 0 & -\pi P & \star \\ \pi B^T PLC & 0 & -\pi B^T PLC & 0 & 0 & 0 & 0 & 0 & 0 & -\pi B^T PB \end{bmatrix} < 0, \quad (5.27)$$

where $\pi = \frac{\bar{\alpha}}{1-\bar{\alpha}}$, $\phi = \frac{\bar{\beta}}{1-\bar{\beta}}$, *and*

$$\begin{aligned} \bar{\Sigma}_{11} =& (A+\Delta A)^T P(A+\Delta A) - (A+\Delta A)^T S^T (SB)^{-1} S(A+\Delta A) - S^T (SB)^{-1} S - P \\ & + \bar{\alpha}Q_2 + (1-\bar{\alpha})Q_1 + \rho I, \\ \bar{\Sigma}_{12} =& S^T (SB)^{-1} S - \bar{\alpha}Q_2 \\ \bar{\Sigma}_{22} =& - S^T (SB)^{-1} S - P + Q_2 + \rho I, \\ \bar{\Sigma}_{33} =& -(1-\bar{\alpha})Q_1 + \bar{\alpha}Q_2 + \rho I. \end{aligned}$$

Hence, using Lemma 2.3 it can be shown that the feasibility of the inequality in (5.27) *is equivalent to that of*

$$\begin{bmatrix} \hat{\Sigma}_{11} & \hat{\Sigma}_{12} & \cdots \\ \hat{\Sigma}_{12}^T & \hat{\Sigma}_{22} & \cdots \\ \vdots & \vdots & \ddots \end{bmatrix} < 0, \quad (5.28)$$

with

$$\begin{aligned} \hat{\Sigma}_{11} =& (A+\Delta A+BF_1)^T P(A+\Delta A+BF_1) - P + (1-\bar{\alpha})Q_1 + F_2^T (B^T PB)F_2 \\ & + F_2^T B^T P + PBF_2 + \bar{\alpha}Q_2 + \rho I, \\ \hat{\Sigma}_{12} =& F_2^T B^T PBF_3 - F_2^T B^T P + PBF_3 - \bar{\alpha}Q_2 \\ \hat{\Sigma}_{22} =& F_3^T (B^T PB)F_3 - F_3^T B^T P - PBF_3 - P + Q_2 + \rho I, \end{aligned} \quad (5.29)$$

where F_1, F_2 *and* F_3 *are auxiliary variables [86] and note that except* $\hat{\Sigma}_{11}$, $\hat{\Sigma}_{12}$ *and* $\hat{\Sigma}_{22}$ *other entries of* (5.28) *are the same as their counterparts in* (5.27). *Thus, using Lemma 3.1,* (5.29) *can be rearranged as*

$$
\begin{aligned}
\hat{\Sigma}_{11} =& [P(A+\Delta A)+BZF_1]^T P^{-1}[P(A+\Delta A)+BZF_1]-P+(1-\bar{\alpha})Q_1 \\
& +F_2^T Z^T B^T P^{-1} BZF_2 + F_2^T Z^T B^T + BZF_2 + \bar{\alpha}Q_2 + \rho I, \\
\hat{\Sigma}_{12} =& F_2^T Z^T B^T P^{-1} BZF_3 - F_2^T Z^T B^T + BZF_3 - \bar{\alpha}Q_2, \\
\hat{\Sigma}_{22} =& F_3^T Z^T B^T P^{-1} BZF_3 - F_3^T Z^T B^T - BZF_3 - P + Q_2 + \rho I,
\end{aligned}
\tag{5.30}
$$

where Z *satisfies* $PB = BZ$. *Using the Schur complement it can be seen that* (5.28) *is implied by the following inequality,*

$$
\left[\begin{array}{cccccccccccc}
\tilde{\Sigma}_{11} & \star & \star & \star & \star & \star & \star & \star & \star & \star & \star & \star \\
\tilde{\Sigma}_{12}^T & \tilde{\Sigma}_{22} & \star & \star & \star & \star & \star & \star & \star & \star & \star & \star \\
BX_2 & BX_3 & -P & \star & \star & \star & \star & \star & \star & \star & \star & \star \\
-\bar{\alpha}Q_2 & \bar{\alpha}Q_2 & 0 & \bar{\Sigma}_{33} & \star & \star & \star & \star & \star & \star & \star & \star \\
0 & 0 & 0 & 0 & -Q_2+\rho I & \star & \star & \star & \star & \star & \star & \star \\
B^T P(\lambda I_n+\Delta A) & B^T P A_\lambda & 0 & 0 & 0 & -B^T PB & \star & \star & \star & \star & \star & \star \\
-\lambda B^T P & B^T(\lambda P - X_4 C) & 0 & 0 & -\pi B^T X_4 C & 0 & -B^T PB & \star & \star & \star & \star & \star \\
-2\phi B^T P A_\lambda & 2\phi B^T P A_\lambda & 0 & 0 & 0 & 0 & 0 & -2\phi B^T PB & \star & \star & \star & \star \\
P\Delta A & PA - X_4 C & 0 & 0 & -\pi X_4 C & 0 & 0 & 0 & -P & \star & \star & \star \\
\pi X_4 C & 0 & 0 & -\pi X_4 C & 0 & 0 & 0 & 0 & 0 & -\pi P & \star & \star \\
\pi B^T X_4 C & 0 & 0 & -\pi B^T X_4 C & 0 & 0 & 0 & 0 & 0 & 0 & -\pi B^T PB & \star \\
P(A+\Delta A)+BX_1 & 0 & 0 & 0 & 0 & 0 & 0 & 0 & 0 & 0 & 0 & -P
\end{array}\right] < 0, \tag{5.31}
$$

where $\tilde{\Sigma}_{11} = -P+(1-\bar{\alpha})Q_1+\bar{\alpha}Q_2+X_2^T B^T + BX_2 + \rho I$, $\tilde{\Sigma}_{12} = -X_2^T B^T + BX_3 - \bar{\alpha}Q_2$, $\tilde{\Sigma}_{22} = -X_3^T B^T - BX_3 - P + Q_2 + \rho I$, $X_1 = ZF_1$, $X_2 = ZF_2$, $X_3 = ZF_3$ *and* $X_4 = PL$. *With the help of Lemma 2.1 and assuming* ΔA_k *satisfies the condition in* (2.2), (5.31) *is also sufficed by the LMI in* (5.15).

Remark 5.3 *Note that the inequality in* (5.27), *which contains the negative quadratic signum terms, cannot easily be converted to an LMI. Let us explain the* *technique that we utilized in the proof of Theorem 5.1 to deal with one of the negative terms. Obviously,* $-\sigma^T(k)(SB)^{-1}\sigma(k)$ *in* $\Delta V(\zeta(k))$ *can be rewritten as*

$$
\begin{aligned}
&-\sigma^T(k)(SB)^{-1}\sigma(k)\\
&=-\begin{bmatrix} x^T(k) & e^T(k)\end{bmatrix}\begin{bmatrix} I \\ -I\end{bmatrix}S^T(SB)^{-1}S\begin{bmatrix} I \\ -I\end{bmatrix}^T\begin{bmatrix} x(k) \\ e(k)\end{bmatrix}.
\end{aligned}
$$

Hence, according to Lemma 2.3, the feasibility of

$$
\tilde{\Psi}-\begin{bmatrix} I \\ -I \\ 0 \\ \vdots \\ 0\end{bmatrix}S^T(SB)^{-1}S\begin{bmatrix} I \\ -I \\ 0 \\ \vdots \\ 0\end{bmatrix}^T<0,
$$

where

$$
\tilde{\Psi}=\Psi+\begin{bmatrix} I \\ -I \\ 0 \\ \vdots \\ 0\end{bmatrix}S^T(SB)^{-1}S\begin{bmatrix} I \\ -I \\ 0 \\ \vdots \\ 0\end{bmatrix}^T,
$$

in which Ψ *is the inequality in* (5.27), *is equivalent to that of*

$$
\tilde{\Psi}+\begin{bmatrix} F_2^T \\ F_3^T \\ 0 \\ \vdots \\ 0\end{bmatrix}(SB)\begin{bmatrix} F_2^T \\ F_3^T \\ 0 \\ \vdots \\ 0\end{bmatrix}^T+\begin{bmatrix} F_2^T \\ F_3^T \\ 0 \\ \vdots \\ 0\end{bmatrix}S\begin{bmatrix} I \\ -I \\ 0 \\ \vdots \\ 0\end{bmatrix}^T+\begin{bmatrix} I \\ -I \\ 0 \\ \vdots \\ 0\end{bmatrix}S^T\begin{bmatrix} F_2^T \\ F_3^T \\ 0 \\ \vdots \\ 0\end{bmatrix}^T<0,
$$

where F_2 *and* F_3 *are auxiliary matrix variables introduced in the proof of Theorem 5.1. As seen this method avoids imposing any conservatism to the problem. However, in order to deal with this problem, [101] uses the trivial inequality* $-\sigma^T(k)(SB)^{-1}\sigma(k)<0$ *and replaces this term with zero. Indeed, this would introduce a significant conservatism to the sliding function design problem.*

The above theorem presents a framework for the design of an ODSMC in order to stabilize the NCS in (5.1). However, it does not present a bound on the system states. The following theorem aims to provide a bound for the obtained stochastic closed-loop system state and the corresponding sliding function.

Theorem 5.2 *In the presence of the exogenous disturbance* $d(k)$, *if the LMI in* (5.15) *is feasible, the bound on the augmented system state* $\zeta(k)$ *is, for the given solution* $P>0, Q_1>0$, $Q_2>0$, $L=P^{-1}X_4$ *and* $\rho>0$ *of* (5.15), *as follows*

$$
\forall \upsilon>0,\ \exists k^\star>0,\ \forall k>k^\star,\ \text{s.t. } \mathbb{E}\left\{\|\zeta(k)\|^2\right\}\leq\frac{\lambda_{\max}(\mathbf{M})}{\hat{\rho}\lambda_1}\gamma+\upsilon, \tag{5.32}
$$

where $\lambda_1=\lambda_{\min}[diag(P,P,Q_1,Q_2,(SB)^{-1})]$, $\mathbf{M}=diag(M_p,Q_1,Q_2)$ *with* $M_P=\begin{bmatrix} P+\mathbf{R} & \mathbf{R} \\ \mathbf{R} & P+\mathbf{R}\end{bmatrix}$ *and* $\mathbf{R}=PB(B^TPB)^{-1}B^TP$, *and* $\gamma=(\tau^2+\kappa^2)\|\Pi+\Sigma_c\|\left\|\mathscr{F}^b\right\|^2$; *here* $\Sigma_c=\begin{bmatrix} \Sigma_{55} & \Sigma_{56} \\ \Sigma_{56}^T & \Sigma_{66}\end{bmatrix}$ *and the scalar variable* $\hat{\rho}>0$ *and matrix variable* $\Pi=\begin{bmatrix} \Pi_{11} & \Pi_{12} \\ \Pi_{12}^T & \Pi_{22}\end{bmatrix}>0$

are obtained from solving the following LMI:

$$\begin{bmatrix} \Omega_1 & \star & \star & \star & \star & \star & \star \\ 0 & (\hat{\rho}-\rho)I & \star & \star & \star & \star & \star \\ 0 & 0 & (\hat{\rho}-\rho)I & \star & \star & \star & \star \\ 0 & 0 & 0 & (\hat{\rho}-\rho)I & \star & \star & \star \\ \bar{\Sigma}_{15}^T & \Sigma_{25}^T & 0 & \Sigma_{45}^T & -\Pi_{11} & \star & \star \\ \Sigma_{16}^T & \Sigma_{26}^T & 0 & 0 & -\Pi_{12}^T & -\Pi_{22} & \star \\ 0 & 0 & 0 & 0 & \Omega_2 & 0 & -\bar{\varepsilon}I \end{bmatrix} < 0, \quad (5.33)$$

where $\bar{\Sigma}_{15} = \lambda PB$, $\Omega_1 = (\hat{\rho}-\rho)I + \bar{\varepsilon}N^T N$, $\Omega_2 = 2M^T PB$, $\bar{\varepsilon} > 0$ *is a scalar variable, and* M *and* N *are known matrices in* (2.2).

Proof 14 *Defining* $v(k) = \left[x^T(k)\ e^T(k)\ x_c^T(k-1)\ e_c^T(k-1)\right]^T$ *and* $\Sigma_v = \begin{bmatrix} \Sigma_{15}^T & \Sigma_{25}^T & \Sigma_{35}^T & \Sigma_{45}^T \\ \Sigma_{16}^T & \Sigma_{26}^T & \Sigma_{36}^T & \Sigma_{46}^T \end{bmatrix}^T$, *according to Lemma 2.2, we have*

$$2v^T(k)\Sigma_v \begin{bmatrix} f_\vartheta(k) \\ \vartheta(k) \end{bmatrix} \leq v^T(k)\Sigma_v\Pi^{-1}\Sigma_v^T v(k) + \begin{bmatrix} f_\vartheta(k) \\ \vartheta(k) \end{bmatrix}^T \Pi \begin{bmatrix} f_\vartheta(k) \\ \vartheta(k) \end{bmatrix}, \quad (5.34)$$

where $\Pi > 0$ *with appropriate dimension. It follows from* (5.25), (5.26) *and* (5.34) *that*

$$\Delta V(\zeta(k)) \leq -v^T(k)[\rho I - \Sigma_v\Pi^{-1}\Sigma_v^T]v(k) + \begin{bmatrix} f_\vartheta(k) \\ \vartheta(k) \end{bmatrix}^T [\Sigma_c + \Pi] \begin{bmatrix} f_\vartheta(k) \\ \vartheta(k) \end{bmatrix}(k), \quad (5.35)$$

in which $\Sigma_c = \begin{bmatrix} \Sigma_{55} & \Sigma_{56} \\ \Sigma_{56}^T & \Sigma_{66} \end{bmatrix}$. *If we choose* $\Pi > 0$ *such that*

$$\hat{\rho}I < \rho I - \Sigma_v\Pi^{-1}\Sigma_v^T, \quad (5.36)$$

where $\rho > \hat{\rho} > 0$, *which is always possible if* $\rho > 0$ *exists, then, it follows from* (5.35) *that*

$$\Delta V(\zeta(k)) \leq -\hat{\rho}v^T(k)v(k) + \begin{bmatrix} f_\vartheta(k) \\ \vartheta(k) \end{bmatrix}^T [\Sigma_c + \Pi] \begin{bmatrix} f_\vartheta(k) \\ \vartheta(k) \end{bmatrix}. \quad (5.37)$$

Moreover, note that as $\sigma(k) = S(x(k) - e(k))$ *we can derive*

$$V(\zeta(k)) = v^T(k)\mathbf{M}v(k), \quad (5.38)$$

where $\mathbf{M} = diag(M_p, Q_1, Q_2)$ *with* $M_P = \begin{bmatrix} P+\mathbf{R} & \mathbf{R} \\ \mathbf{R} & P+\mathbf{R} \end{bmatrix}$ *and* $\mathbf{R} = PB(B^T PB)^{-1}B^T P$, *hence,*

$$\lambda_{\min}(\mathbf{M})\|v(k)\|^2 \leq V(\zeta(k)) \leq \lambda_{\max}(\mathbf{M})\|v(k)\|^2. \quad (5.39)$$

Additionally, it is known that

$$\lambda_1\|\zeta(k)\|^2 \leq V(\zeta(k)) \leq \lambda_2\|\zeta(k)\|^2. \quad (5.40)$$

where $\lambda_1 = \lambda_{\min}(diag(P,P,Q_1,Q_2,(SB)^{-1}))$ *and* $\lambda_2 = \lambda_{\max}(diag(P,P,Q_1,Q_2,(SB)^{-1}))$. *Hence, from* (5.37) *and* (5.39) *one can derive that*

$$\Delta V(\zeta(k)) \leq -\frac{\hat{\rho}}{\lambda_{\max}(\mathbf{M})}V(\zeta(k)) + \gamma, \quad (5.41)$$

where $\gamma = (\tau^2+\kappa^2)\|\Pi+\Sigma_c\|\left\|\mathscr{F}^b\right\|^2$. *Note that from* (5.26) *it can simply be written that* $\forall\ \nu(k) \neq 0$

$$\nu^T(k)\Xi\nu(k) = \mathbb{E}\left\{V(\zeta(k+1))\big|_{\left[\begin{smallmatrix} f_\vartheta(k) \\ \vartheta(k)\end{smallmatrix}\right]=0}\Big|\zeta(k)\right\} - V(\zeta(k))$$
$$< -\rho\nu^T(k)\nu(k). \tag{5.42}$$

It is known that $\mathbb{E}\{V(\zeta(k+1))\big|_{\left[\begin{smallmatrix} f_\vartheta(k) \\ \vartheta(k)\end{smallmatrix}\right]=0}\big|\zeta(k)\} > 0$, *and thus, from* (5.42) *and* (5.39), *it can be claimed that* $\rho < \lambda_{\max}(\mathbf{M})$. *Hence,*

$$\frac{\hat{\rho}}{\lambda_{\max}(\mathbf{M})} < 1.$$

Finally, from Lemma 2.4 and (5.41), *we can find the bound given in* (5.32).

Furthermore, to find $\Pi > 0$ *in* (5.36), *defining* $\Pi = \left[\begin{smallmatrix} \Pi_{11} & \Pi_{12} \\ \Pi_{12}^T & \Pi_{22}\end{smallmatrix}\right]$, *we can rewrite this inequality, for given* $P > 0$, $Q_1 > 0$, $Q_2 > 0$, L *and* $\rho > 0$, *as*

$$\begin{bmatrix} (\hat{\rho}-\rho)I & \star & \star & \star & \star & \star \\ 0 & (\hat{\rho}-\rho)I & \star & \star & \star & \star \\ 0 & 0 & (\hat{\rho}-\rho)I & \star & \star & \star \\ 0 & 0 & 0 & (\hat{\rho}-\rho)I & \star & \star \\ \Sigma_{15}^T & \Sigma_{25}^T & 0 & \Sigma_{45}^T & -\Pi_{11} & \star \\ \Sigma_{16}^T & \Sigma_{26}^T & 0 & 0 & -\Pi_{12}^T & -\Pi_{22} \end{bmatrix} < 0. \tag{5.43}$$

Using Lemma 2.1, (5.43) *can be sufficed by the LMI* (5.33).

Remark 5.4 *Notice that, in the framework in [101], the cross terms between the system states and the exogenous disturbance have increased the conservatism of the final obtained LMI condition. It has been shown here that the mentioned cross terms should not influence the feasibility region of the final LMI condition* (5.15). *Besides, instead of removing the cross terms between* $x(k)$, $e(k)$, $x_c(k-1)$ *and* $e_c(k-1)$ *through several inequalities which was implemented in [101] for* $x(k)$ *and* $x(k-1)$, *we let the cross terms to be in the original form. This would also widen the feasible region of the LMI condition presented in this work.*

The solution of the LMI in (5.33) does not have direct influence on the controller design and the actual bound on the system state and/or sliding function, however, these parameters would lead us to determine a more accurate bound. Therefore, to obtain the minimum value of the bound in (5.32), the LMIs in (5.15) and (5.33) could be solved subject to a specific criteria. This issue is beyond the scope of this book and remains for future work.

Remark 5.5 *The main idea in the proof of Theorems 5.1, 5.2 can be summarized as follows. First, we tried to find*

$$\Delta V(\zeta(k)) \triangleq \mathbb{E}\{V(\zeta(k+1))\big|\zeta(k)\} - V(\zeta(k))$$
$$= \varpi^T(k)\left[\Sigma_{ij}\right]_{6\times 6}\varpi(k)$$

where $\zeta(k)$, $\varpi(k)$ and Σ_{ij}, $i, j = 1, \cdots, 6$ are defined in the proof of Theorem 5.1. Then, we let

$$[\Sigma_{ij}]_{6\times6} = \begin{bmatrix} [\Sigma_{ij}]_{4\times4} & \Sigma_v \\ \Sigma_v^T & \Sigma_c \end{bmatrix},$$

where $\Sigma_c = \begin{bmatrix} \Sigma_{55} & \Sigma_{56} \\ \Sigma_{56}^T & \Sigma_{66} \end{bmatrix}$ *and* $\Sigma_v = \begin{bmatrix} \Sigma_{15}^T & \Sigma_{25}^T & \Sigma_{35}^T & \Sigma_{45}^T \\ \Sigma_{16}^T & \Sigma_{26}^T & \Sigma_{36}^T & \Sigma_{46}^T \end{bmatrix}^T$. *Now, the switching function matrix $S = B^T P$, and the observer gain L can be obtained through solving the LMI condition (5.15), which is derived by analyzing the feasibility of the following inequality:*

$$[\Sigma_{ij}]_{4\times4} < -\rho I,$$

where $\rho > 0$ is a scalar variable, with the aid of several convexification techniques. The obtained observer-based linear control law in (5.12) ensures the mean square stability of the augmented closed-loop system (in the absence of the exogenous disturbances). Theorem 5.2, then, characterizes the boundedness of the augmented closed-loop system when the exogenous disturbances are present, using an innovative scheme. Please also notice that the basic idea of the proof of this theorem comes from Lemma 2.4.

5.4 Numerical Examples

Two numerical examples are presented here in order to evaluate the effectiveness of the proposed ODSMC. All the LMI optimization problems are solved by YALMIP [90] as the interface and SDPT3 [136] as the solver.

5.4.1 Example 1

Consider the system (5.1) with the following parameters:

$$A = \begin{bmatrix} 0.25 & 0 & 0.28 \\ 0 & 1.00 & -0.20 \\ 0.50 & 0 & 0.40 \end{bmatrix}, B = \begin{bmatrix} 1 & 0.2 \\ 0.5 & 0 \\ 0 & 1 \end{bmatrix}, C = \begin{bmatrix} 1 & 0 & 1 \\ 0.5 & 1 & 0.3 \end{bmatrix},$$

$$M = \begin{bmatrix} 0 & 0.1 & -0.1 \\ 0 & 0.05 & 0.02 \\ 0 & 0 & -0.1 \end{bmatrix}, N = \begin{bmatrix} 0 & 0.15 & 0.01 \\ 0.01 & -0.02 & 0 \\ 0.01 & 0 & 0.1 \end{bmatrix},$$

$$R(k) = \text{diag}(0.3\sin(k), 0.9\sin(k), 0.6\cos(k)),\ \bar{\alpha} = 0.15,\ \bar{\beta} = 0.2.$$

As can be seen, the open-loop system is unstable. Suppose

$$d(k) = \begin{bmatrix} 0.05 \\ 0.1 \end{bmatrix} \sin(\frac{k}{10}).$$

Solving the LMI in (5.15) gives the following results:

$$P = \begin{bmatrix} 34.3162 & 0.0546 & -1.4185 \\ 0.0546 & 34.5799 & -0.6111 \\ -1.4185 & -0.6111 & 33.5812 \end{bmatrix}, \; Q_1 = \begin{bmatrix} 18.3586 & 3.0846 & -3.2345 \\ 3.0846 & 3.3992 & 1.9403 \\ -3.2345 & 1.9403 & 18.0901 \end{bmatrix}, \tag{5.44}$$

$$Q_2 = \begin{bmatrix} 11.4218 & 3.1738 & -1.1569 \\ 3.1738 & 3.0348 & 1.0398 \\ -1.1569 & 1.0398 & 14.1352 \end{bmatrix}, \; S = \begin{bmatrix} 34.3435 & 17.3446 & -1.7240 \\ 5.4448 & -0.6002 & 33.2975 \end{bmatrix},$$

$$L = \begin{bmatrix} 0.1605 & -0.0039 \\ -0.1422 & 0.2904 \\ 0.1875 & 0.0535 \end{bmatrix}, \; \rho = 0.2083, \; \varepsilon = 24.5523.$$

The component $\vartheta(k)$ in (5.10) is assumed to be as

$$\vartheta(k) = \mathscr{F}^a + \eta \mathrm{diag}(\mathscr{F}^b)\mathrm{sgn}(S\hat{x}(k)), \tag{5.45}$$

where $\mathrm{diag}(\mathscr{F}^b) = \mathrm{diag}(d_1^b, \cdots, d_m^b)$ and $\eta > 0$ is a scalar. Hence, using P, $\mathscr{F}^a = \begin{bmatrix} 0 & 0 \end{bmatrix}^T$, $\mathscr{F}^b = \begin{bmatrix} 0.05 & 0.1 \end{bmatrix}^T$, $\lambda = 0.3$ and $\eta = 0.01$ the control law given in (5.10) and (5.45) is obtained. The results of applying this controller to the system (5.1) are shown in Figs. 5.3-5.6. Here, the initial state is assumed to be $x(0) = \begin{bmatrix} 2 & 0 & -3 \end{bmatrix}^T$. Bernoulli sequences $\alpha(k)$ and $\beta(k)$ are depicted in Fig. 5.7. It can be seen that the proposed ODSMC law successfully drives the state trajectories toward the ideal sliding surface and keeps them in a boundary layer around the sliding surface thereafter.

5.5 Conclusions

This chapter proposes a robust observer-based discrete-time sliding mode control by utilizing only output signals for the networked systems involving random measuring and actuating consecutive packet losses. We have exploited Bernoulli random binary distribution to model the consecutive data packet dropouts. Besides, the proposed method, achieved with the aid of an LMI scheme, forms a unified framework for the robust ODSMC design. Furthermore, it has reduced the conservatism of the existing methods in the literature. For analyzing the ultimate boundedness of the derived closed-loop system, the notion of exponentially mean square stability has been utilized. Numerical examples have been presented to show the effectiveness of the proposed scheme.

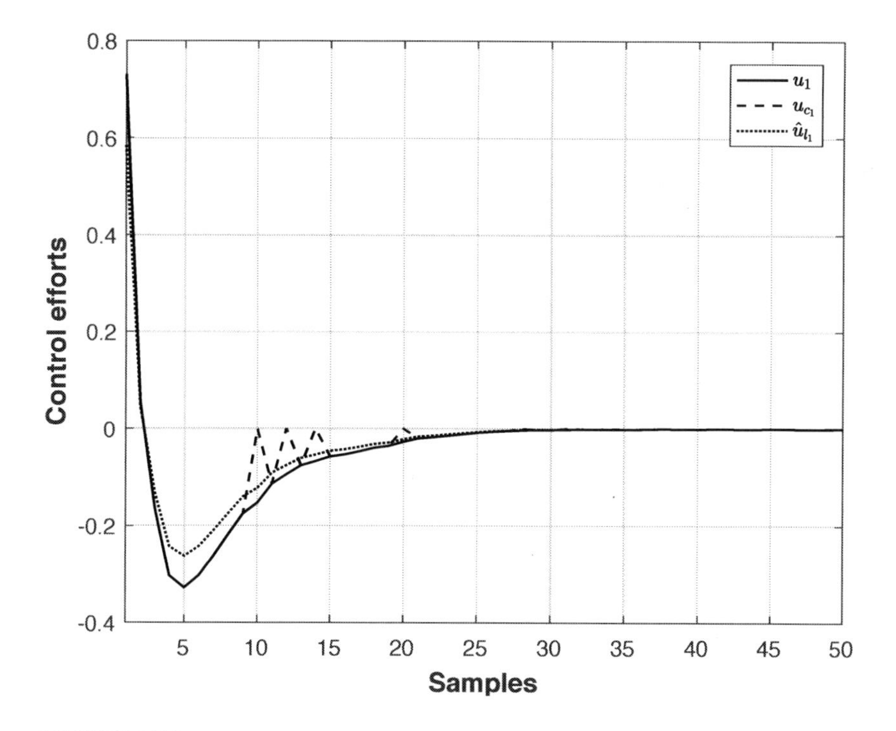

FIGURE 5.2
Control effort u_1.

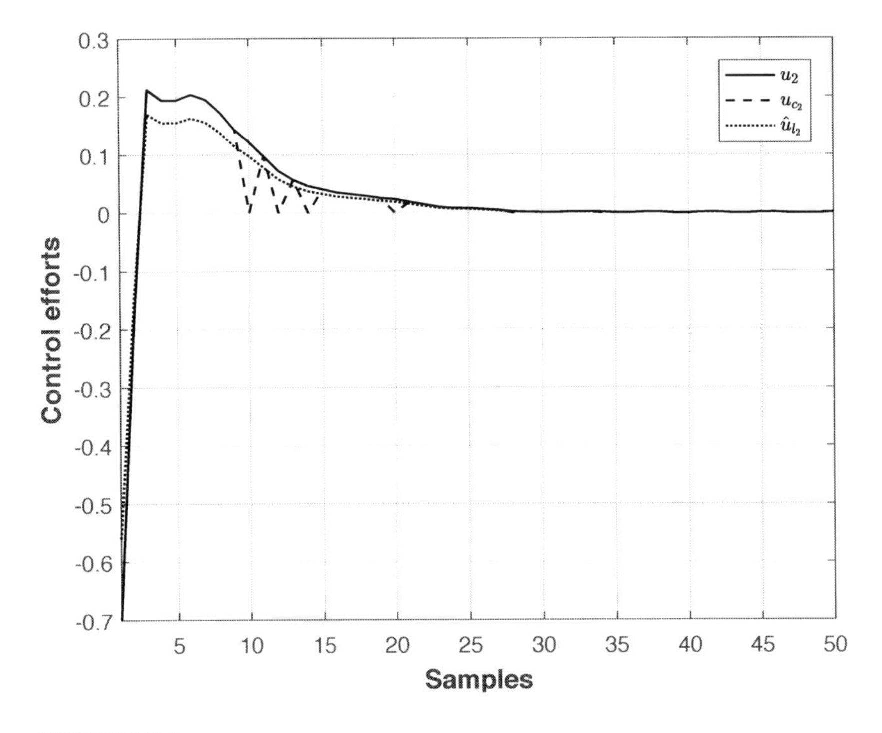

FIGURE 5.3
Control effort u_2.

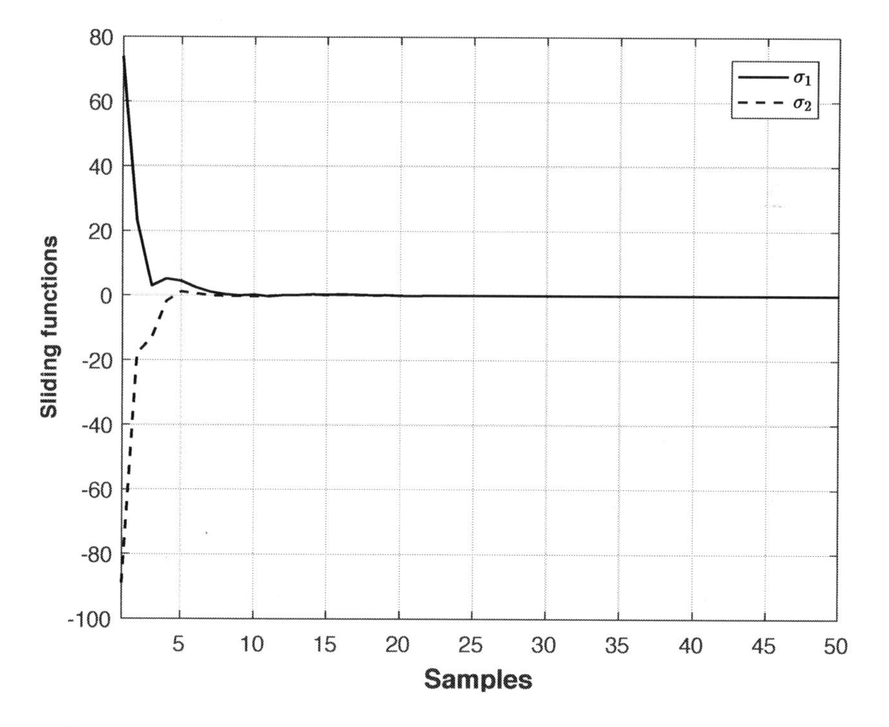

FIGURE 5.4
Trajectories of the sliding function.

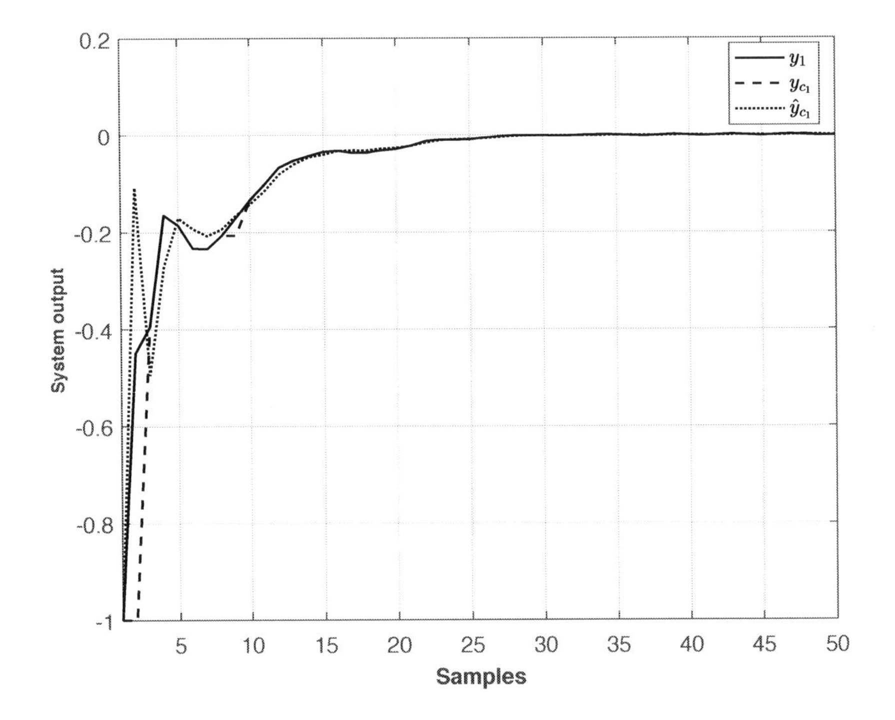

FIGURE 5.5
Trajectories of the system output and its estimate.

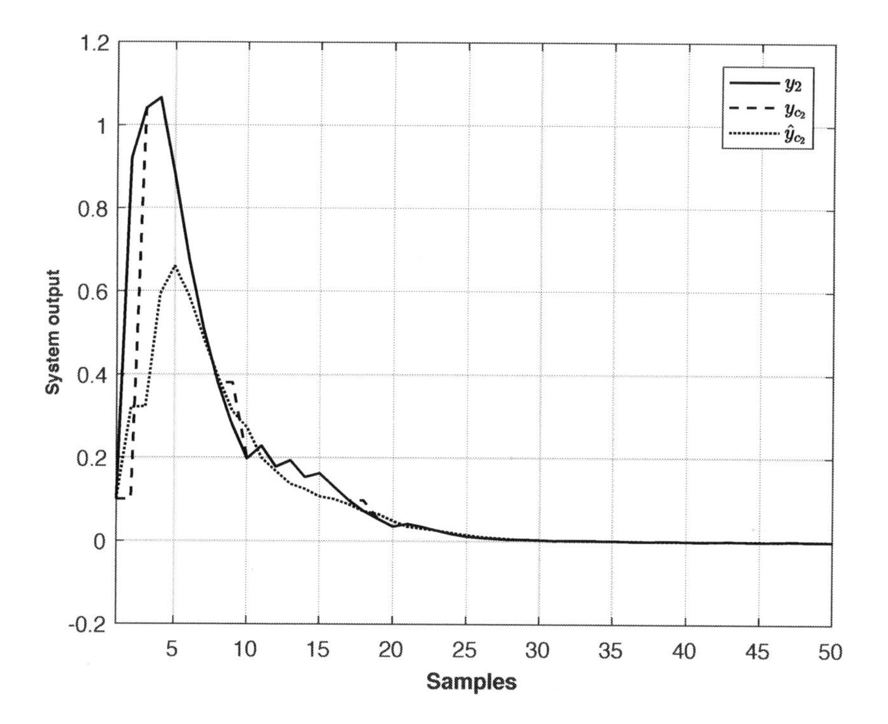

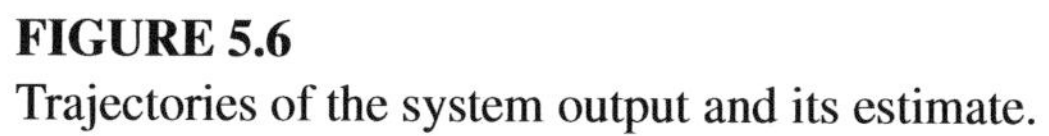

FIGURE 5.6
Trajectories of the system output and its estimate.

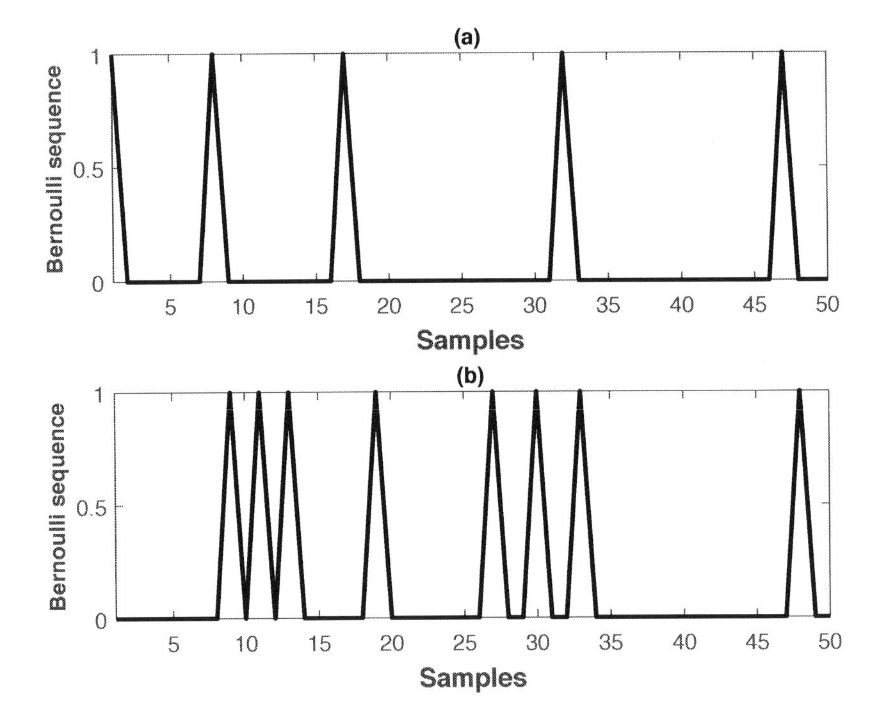

FIGURE 5.7
Bernoulli sequences (a) $\alpha(k)$, (b) $\beta(k)$.

Part III

Sparse Sliding Mode Control for Large Scale NCSs

6

Sparse Observer-based discrete-time SMC for NCSs

CONTENTS

6.1	Introduction	113
6.2	Problem Formulation and Preliminaries	116
	6.2.1 Preliminaries	117
	6.2.2 State and disturbance observer	119
6.3	Spatially Decentralized Sliding Mode Control	120
6.4	Stability Analysis	122
6.5	Sparsifying the Control Network Structure	127
6.6	Numerical Examples	129
	6.6.1 Example 1	129
	6.6.2 Example 2	130
	6.6.3 Example 3	138
6.7	Conclusions	138

Abstract– This chapter proposes a framework for the design of sparsely distributed output feedback discrete-time sliding mode control (ODSMC) for interconnected systems. In fact, the main objective here is to develop an observer based discrete-time sliding mode controller employing a sparsely distributed control network structure in which local controllers exploit some other sub-systems' information as well as its own local information. As the local controllers/observers have access to some other sub-systems' states, the control performance will be improved and the applicability region will be widened compared to the decentralized structure. As the first step, a stability condition is derived for the overall closed-loop system obtained from applying ODSMC to the underlying interconnected system, by assuming an a priori known structure for the control/observer network. The developed LMI based controller design scheme provides the possibility to employ different information patterns such as fully distributed, sparsely distributed and decentralized patterns. In the second step, a methodology is proposed to identify a sparse control/observer network structure with the least possible number of communication links that satisfies the stability condition given in the first step.

6.1 Introduction

Spatially distributed control systems which exploit communication networks in their loop have been also regarded as networked control systems (NCS). Utilizing a centralized control scheme in NCSs which requires the central controller to have access to the states of all subsystems' plants is not practical as it needs a larger and more costly control network. On the other hand, alternatively, decentralized or distributed control architectures have been proposed and used in the literature [155, 139, 26, 63]. The general idea behind the decentralized control scheme is to use only the local state information in order to control the subsystems and thus there is no control network. This can be effective only when the interconnections between the subsystems are not strong [150, 125]. When the interconnections are strong, utilizing the distributed control frameworks has been considered to ensure stability of the overall system [114]. In this strategy, each subsystem can exploit the local state as well as the state of some other subsystems. Hence, compared to the decentralized control scheme, the distributed control scheme can ensure stability of the overall system in the presence of stronger subsystem interconnections [140]. Meanwhile, it also has lower complexity and improved computational aspects compared to the centralized control scheme.

While a large number of investigations in the control systems literature focus on the analysis of continuous-time systems, more and more practising control engineers implement the control laws using micro-processors [41]. The controllers can either be carried out from continuous-time representations using fast sampling ideas, or the continuous-time controllers can be converted to their discrete-time representations. However, the choice of high sampling rate, which nearly approximates continuous-time, may not always be possible [83]. Alternatively, discrete-time controllers can be designed directly from a discrete-time representation of the plant. One thread of the literature develops discrete-time controllers to stabilize discrete-time uncertain linear systems with bounded uncertainties. A great deal of the work in this field considered state feedback control laws based on Lyapunov ideas [41]. Alternatively, the idea of discrete-time sliding mode control (DSMC) has been proposed in the literature; see [83] for more information, which is significantly different from its continuous-time counterpart. The results presented in e.g. [83] demonstrate that an appropriate choice of sliding surface, used with the *equivalent control*, can ensure a bounded motion about the surface in the presence of bounded matched uncertainty. Notice also that from this viewpoint, the DSMC problem can be seen as a robust optimal control problem and is related to discrete-time Lyapunov min-max problems [83]. The problem is to select, among all possible feedback controllers, the feedback gain that minimizes the worst case effect of the uncertainty on the Lyapunov difference function [83]. Moreover, the discrete-time equivalent control law can be considered as a solution of the discrete-time LQR problem under the assumption of *cheap control*; that is, no penalty is assigned to the control effort in the cost function. On the other hand, much of the work performed on the design of sparsely distributed controller/observer considered static output feedback control laws. Alternatively, the objective of this chapter is to extend this idea to the field of DSMC problems.

An outstanding research implemented on the sliding mode control (SMC) has been decentralized SMC for the large-scale interconnected systems; see [144, 145, 112, 92] and the references therein. Furthermore, in the literature the distributed SMC has received less attention and hence it requires more investigation. This chapter first explores the problem of designing a sparse DSMC network for a given plant network with arbitrary topology. To do so, this chapter considers an a priori control network topology which is a subset of an underlying dynamics network and provides a methodology to stabilize the underlying dynamics utilizing a (sparse) distributed observer and controller network. It is also assumed that communication networks do not have any data packet loss, bandwidth limitation, or network delays. We will show that the proposed observer-based DSMC has this ability to cover all the cases such as fully decentralized, fully distributed, and sparsely distributed topologies. Exploiting a sparse structure for the control network which is a subset of the dynamics structure is crucial in the control system for (very) large scale systems, for instance, the smart grids [4].

While addressing the problem of designing a sparsely distributed observer-based DSMC strategy with assuming a priori topology for the control network is important, another arising problem will be providing a method to find a sparse control network structure. Taking a broad look at the issue, one may resort to finding the sparsest control/observer structure, which is necessarily a subset of the dynamics network, but still can ensure the stability of the interconnected system. This issue has been investigated in e.g. [88] in order to find the suboptimal controllers that minimize a special objective function considering the sparsity of the feedback gain. However, this may result in a non-convex condition. Furthermore, [114] considers the problem of finding the sparsest control/observer network that satisfies a set of stability conditions, obtained through a Lyapunov direct scheme. In this chapter, as the second step, we will search for a sparse control/observer network structure with the least possible number of links that can satisfy the given stability condition. To this end, a heuristic iterative algorithm will be proposed, distinguishing itself from a trial-and-error process which requires checking all the possible structures.

Disturbance observer-based control strategies have been exploited in different fields in the literature and have been successfully implemented for different aims; see e.g. [87]. The idea of using a disturbance estimator in the DSMC has been developed in [133] in order to reduce the ultimate bound on the discrete-time system state. However, the disturbance estimator in [133] has been designed for the cases when the system states are fully available and the system does not involve unmatched uncertainties. A framework has been proposed in [32] by exploiting output information only for discrete-time MIMO systems involving unmatched exogenous disturbances and without unmatched uncertainties. Indeed, the idea is to use an integral term of the estimation output error, in addition to the well-known Luenberger observer which observes the system state with a proportional loop, to make more degrees of freedom. This matter is referred to as *proportional integral observer* (PIO) in the literature [32]. The distributed output feedback DSMC (ODSMC), presented in this chapter, utilizes a disturbance observer in order to deal with the influences of the exogenous disturbances on the boundary layer thickness. This sparsely distributed ODSMC is designed by means of an LMI scheme.

The rest of this chapter is as follows. Section 6.2 describes the problem formulation and preliminaries. Section 6.3 presents the sparse distributed sliding mode control and the sparse full-order observer. Section 6.4 shows the LMI-based Lyapunov stability scheme utilized for the design of robust sliding surface, state and disturbance observer gains concurrently. Section 6.5 explores a control network structure with the minimum number of links that satisfy the LMI stability condition derived in Section 6.4. Effectiveness of the proposed sparse distributed ODSMC is studied by numerical examples in Section 6.6. Finally, Section 6.7 concludes this chapter.

6.2 Problem Formulation and Preliminaries

Consider a large scale networked system consisting of h sub-systems:

$$\begin{cases} x_i(k+1) = [A_{ii} + \Delta A_{ii}(k)]x_i(k) + B_i[u_i(k) + f_i(k)] \\ \qquad\qquad + \sum\limits_{j=1,\ j\neq i}^{h} [A_{ij} + \Delta A_{ij}(k)]x_j(k), \\ y_i(k) = C_i x_i(k), \quad i = 1, \cdots, h, \end{cases} \tag{6.1}$$

where $x_i \in \mathbb{R}^{n_i}$, $y_i \in \mathbb{R}^{p_i}$ and $u_i \in \mathbb{R}^{m_i}$ are the state vector, output vector and control input vector of the i-th sub-system, respectively. The matrices in (6.1) are constant and of appropriate dimensions. Without loss of generality, it is assumed that $m_i \leq p_i \leq n_i$, $\mathrm{rank}(B_i) = m_i$, $\mathrm{rank}(C_i) = p_i$. The term $\Delta A_{ii}(k)$ denotes the uncertainty of i-th sub-system and $\sum_{j=1, j\neq i}^{h} A_{ij}x_j(k)$, $\sum_{j=1, j\neq i}^{h} \Delta A_{ij}(k)x_j(k)$ are, respectively, a known interconnection and an uncertain interconnection of the i-th sub-system. We also assume

$$\Delta A_{ij}(k) = M_{ij}R_{ij}(k)N_{ij}, \tag{6.2}$$

where M_{ij} and N_{ij} are known matrices and $R_{ij}(k)$ is an unknown time varying matrix satisfying $R_{ij}^T(k)R_{ij}(k) \leq I$. $f_i(k)$ is the matched external disturbances of the i-th sub-system with known bound.

Notice that unlike most of the literature of SMC that considers the systems involving perturbations satisfying a matching condition, in order to make the problem closer to practical cases and improve the generality of the controller synthesis problem, in this chapter, we study the systems involving mismatched uncertainties. Indeed, the switching surface design problem, and further, the problem of synthesizing the linear part of the SMC can be solved via LMI methods such that the obtained linear controller, in the absence of non-vanishing matched exogenous disturbances, can quadratically stabilize the system for all admissible mismatched norm-bounded time-varying uncertainties

Define

$$x(k) := \mathrm{col}(x_i(k))_{i=1}^h,\ u(k) := \mathrm{col}(u_i(k))_{i=1}^h,$$
$$y(k) := \mathrm{col}(y_i(k))_{i=1}^h,\ f(k) := \mathrm{col}(f_i(k))_{i=1}^h, \tag{6.3}$$

and

$$A := [A_{ij}]_{h\times h},\ \Delta A(k) := [\Delta A_{ij}(k)]_{h\times h},$$
$$B := \mathrm{diag}[B_i]_{i=1}^h,\ C := \mathrm{diag}[C_i]_{i=1}^h. \tag{6.4}$$

Besides, the uncertainty matrix for overall system can be rearranged as:

$$\Delta A(k) = \sum_{i=1}^{h} M_i R_i(k) N_i, \tag{6.5}$$

in which

$$M_i = [M_{rj}^i]_{h\times h},\ M_{rj}^i = \begin{cases} M_{ij} & \text{if } r = i, \\ 0 & \text{otherwise,} \end{cases}$$
$$R_i(k) = \mathrm{diag}(R_{i1}(k), \cdots, R_{ih}(k)),$$
$$N_i = \mathrm{diag}(N_{i1}, \cdots, N_{ih}).$$

Using (6.1), (6.3) and (6.4), the system in (6.1) at a network level can be written

$$\begin{cases} x(k+1) = [A + \Delta A(k)]x(k) + B[u(k) + f(k)] \\ y(k) = Cx(k). \end{cases} \tag{6.6}$$

Note that the proposed method is not restrictive to this ideal system, but can be readily extended to e.g. the system with time-delay and package losses; see [6] for our relevant work in the same DSMC framework.

6.2.1 Preliminaries

Definition 6.1 *A matrix is said to be a structure matrix if its elements are either 0 or 1. The structure matrix of a block matrix* $Y = [Y_{ij}]_{h\times h}$ *with* $Y_{ij} \in \mathbb{R}^{r_i \times s_j}$ *is* $\mathsf{S}(Y) \triangleq [s_{ij}]_{h\times h}$ *with*

$$s_{ij} = \begin{cases} 0 & \textit{if } Y_{ij} = 0,\ i \neq j \\ 1 & \textit{otherwise.} \end{cases}$$

Notice that the structure matrix defined above is somewhat similar to the well-known *adjacency* matrix in *graph theory*; see [55]. However, unlike the *adjacency* matrix, the diagonal entries in the structure matrix will be assumed to be 1.

Definition 6.2 *Two matrices* Y_1 *and* Y_2 *are said to have the same structure if* $\mathsf{S}(Y_1) = \mathsf{S}(Y_2)$.

Definition 6.3 *The matrix* Y_1 *with* $\mathsf{S}(Y_1) \triangleq [s_{ij}^1]_{h\times h}$ *is said to be structurally subset of* Y_2 *with* $\mathsf{S}(Y_2) \triangleq [s_{ij}^2]_{h\times h}$ *while* $s_{ij}^2 - s_{ij}^1 \geq 0$. *We denote this as* $\mathsf{S}(Y_1) \subseteq \mathsf{S}(Y_2)$.

Lemma 6.1 *Consider* $0 < W = \begin{bmatrix} diag[\bar{W}_i]_{i=1}^h & diag[\hat{W}_i]_{i=1}^h \\ diag[\hat{W}_i]_{i=1}^{h^T} & diag[\tilde{W}_i]_{i=1}^h \end{bmatrix} \in \mathbb{R}^{(m+n)\times(m+n)}$, *with* $\bar{W}_i \in \mathbb{R}^{n_i\times n_i}$, $\tilde{W}_i \in \mathbb{R}^{m_i\times m_i}$, $\hat{W}_i \in \mathbb{R}^{n_i\times m_i}$. *We have*

$$\mathsf{S}(W) = \mathsf{S}(W^{-1}).$$

Proof 15 *Since* $W = \begin{bmatrix} \Theta_1 & \Theta_2 \\ \Theta_2^T & \Theta_3 \end{bmatrix} > 0$, *it is known that* $W^{-1} > 0$. *Suppose*

$$W^{-1} = \begin{bmatrix} \bar{\Theta}_1 & \bar{\Theta}_2 \\ \bar{\Theta}_2^T & \bar{\Theta}_3 \end{bmatrix}.$$

It is then easy to show that [108]

$$\begin{aligned} \bar{\Theta}_1 &= (\Theta_1 - \Theta_2\Theta_3^{-1}\Theta_2^T)^{-1}, \\ \bar{\Theta}_3 &= (\Theta_3 - \Theta_2^T\Theta_1^{-1}\Theta_2)^{-1}, \\ \bar{\Theta}_2 &= -\Theta_1^{-1}\Theta_2(\Theta_3 - \Theta_2^T\Theta_1^{-1}\Theta_2)^{-1}, \end{aligned}$$

are block-diagonal matrices and thus $\mathsf{S}(W) = \mathsf{S}(W^{-1})$. *Notice that according to the Schur complement* $\Theta_1 - \Theta_2\Theta_3^{-1}\Theta_2^T$ *and* $\Theta_3 - \Theta_2^T\Theta_1^{-1}\Theta_2$ *are invertible.*

Lemma 6.2 *Consider the matrix* $0 < W \in \mathbb{R}^{(m+n)\times(m+n)}$ *given in Lemma 6.1, for any* $Y = \begin{bmatrix} Y_1 \\ Y_2 \end{bmatrix}$, *where* $Y_1 \in \mathbb{R}^{n\times p}$ *and* $Y_2 \in \mathbb{R}^{m\times p}$, *while* $\begin{bmatrix} J_1 \\ J_2 \end{bmatrix} = W^{-1}Y$ *and* $\mathsf{S}(Y_1) \subseteq \Gamma$, $\mathsf{S}(Y_2) \subseteq \Gamma$ *we have*

$$\begin{aligned} &\mathsf{S}(J_1) \subseteq \Gamma, \\ &\mathsf{S}(J_2) \subseteq \Gamma. \end{aligned}$$

Proof 16 *Due to the simplicity, we omit the proof here.*

Definition 6.4 *The overall system* (6.6) *is said to be structurally controllable with respect to the structure matrix* $\Gamma = [\gamma_{ij}]_{h\times h}$ *if there exists* $K = [K_{ij}]_{h\times h}$ *with* $\mathsf{S}(K) \subseteq \Gamma$ *such that the modes of* $A - BK$ *are arbitrarily assignable.*

Definition 6.5 *The overall system* (6.6) *is said to be structurally observable with respect to the structure matrix* $\Gamma = [\gamma_{ij}]_{h\times h}$ *if there exists* $L = [L_{ij}]_{h\times h}$ *with* $\mathsf{S}(L) \subseteq \Gamma$ *such that* $\Gamma \circ A - LC$ *is Hurwitz, where* $\Gamma \circ A = [\gamma_{ij}A_{ij}]_{h\times h}$.

Assumption 6.1 *The matrix triple (A, B, C) in* (6.6) *is structurally controllable and observable with respect to the given structure matrix* $\Gamma = [\gamma_{ij}]_{h\times h}$.

The following assumption and lemmas are required in the sequel of this chapter.

Assumption 6.2 ([32]) *The matrices A, B and C in the system* (6.6) *and the structure matrix* Γ *satisfy*

$$rank\left(\begin{bmatrix} \Gamma\circ A - I_n & B \\ C & 0 \end{bmatrix}\right) = n+m.$$

Notice that the above assumption is equivalent to not having transmission zero at 1.

Assumption 6.3 *The exogenous disturbance $f_i(k)$ in* (6.1) *satisfies the Lipschitz continuity condition:*

$$\left\|\tilde{f}_i(k)\right\| \leq \tilde{L}_i T_s, \quad \forall k \geq 0, \tag{6.7}$$

where $\tilde{f}_i(k) = f_i(k) - f_i(k-1)$, $\tilde{L}_i > 0$ denotes the Lipschitz constant and T_s is the sampling time.

Here, it is supposed that $\tilde{L}_i$ is small. To this end, the sampling rate of the discrete signal processing system is assumed to be large enough compared to the maximum frequency component of the exogenous disturbance $f_i(k)$.

6.2.2 State and disturbance observer

It is assumed in this chapter that some additional state estimates from other subsystems are utilized to improve the performance of the control loop. This idea is different from the purely decentralized controller and will lead to a *partially decentralized* or distributed control structure. In this chapter, we use the following estimation scheme to provide the i-th local controller by the system state information and disturbance estimate,

$$\begin{cases} \hat{x}_i(k+1) = A_{ii}\hat{x}_i(k) + B_i u_i(k) + \sum\limits_{j=1,\ j\neq i}^{j=h} \gamma_{ij} A_{ij} \hat{x}_j(k) \\ \qquad + \sum\limits_{j=1}^{j=h} \gamma_{ij} L_{ij}[y_j(k) - \hat{y}_j(k)] + B_i \hat{f}_i(k) \\ \hat{f}_i(k+1) = \hat{f}_i(k) + \sum\limits_{j=1}^{j=h} \gamma_{ij} D_{ij}[y_j(k) - \hat{y}_j(k)] \\ \hat{y}_i(k) = C_i \hat{x}_i(k), \end{cases} \tag{6.8}$$

where $\hat{x}_i(k) \in \mathbb{R}^{n_i}$ is the state estimate of the i-th sub-system in (6.1), $\hat{y}_i(k) \in \mathbb{R}^{p_i}$ is the observer output, $L_{ij} \in \mathbb{R}^{n_i \times p_j}$ and $D_{ij} \in \mathbb{R}^{m_i \times p_j}$ are the local observer gains for the state and disturbance respectively. Here γ_{ij} denotes the availability of communication links among subsystems in the controller and observer design, that is, $\gamma_{ij} = 1$ if ij-th link exists in the control/observer network and $\gamma_{ij} = 0$ otherwise. Then the overall estimator is

$$\begin{cases} \hat{x}(k+1) = \Gamma \circ A\hat{x}(k) + Bu(k) + L_s[y(k) - \hat{y}(k)] + B\hat{f}(k) \\ \hat{f}(k+1) = \hat{f}(k) + D_s[y(k) - \hat{y}(k)] \\ \hat{y}(k) = C\hat{x}(k), \end{cases} \tag{6.9}$$

where $L_s := \Gamma \circ L$ with $L = [L_{ij}]_{h\times h}$, $D_s := \Gamma \circ D$ with $D = [D_{ij}]_{h\times h}$ and $\Gamma = [\gamma_{ij}]_{h\times h}$.

6.3 Spatially Decentralized Sliding Mode Control

Consider the following linear sliding function

$$\sigma(k) = Sx(k), \tag{6.10}$$

where $\sigma(k) := \mathrm{col}(\sigma_i(k))_{i=1}^h$ and the block diagonal matrix $S := \mathrm{diag}[S_i]_{i=1}^h$ will be designed later such that SB is nonsingular.

During the ideal sliding motion the sliding function satisfies:

$$\sigma(k) = 0, \quad \forall k > k_s, \tag{6.11}$$

where $k_s > 0$ denotes the time that sliding motion starts. One may obtain from (6.6) and (6.10) that

$$\sigma(k+1) = S[A + \Delta A]x(k) + SB[u(k) + f(k)]. \tag{6.12}$$

Remark 6.1 *A question which would arise here is why the sliding function* (6.10) *is assumed to be in the state space rather than state estimate space or estimation error space, while the overall system output* $y(k)$ *is only available. In fact, since this sliding function would not be used in the variable structure discontinuous component of the ODSMC, the sliding surface is not required to be designed by utilizing known information of the system. Instead, it is only required to be ensured that the system state trajectories could be steered into a boundary layer around the sliding surface and be kept there thereafter. This can basically be regarded as the key feature of the ODSMC presented in this chapter for NCSs. This can also lead to a considerable extension to the applicable region of the framework given in this chapter compared to the existing literature for the continuous-time counterpart. The same manner can be seen in [83] for the static ODSMC.*

The controller is assumed to have the following structure:

$$\begin{aligned} u_i(k) = -(S_iB_i)^{-1}[(S_iA_{ii} - \Phi_iS_i)\hat{x}_i(k) \\ + S_i \sum_{j=1,\, j\neq i}^{j=h} \gamma_{ij}A_{ij}\hat{x}_j(k)] - \hat{f}_i(k), \end{aligned} \tag{6.13}$$

where $\Phi_i \in \mathbb{R}^{m_i \times m_i}$ is a stable matrix and aims to govern the convergence rate of the state driven into a boundary layer around the ideal sliding surface. Here, similar to [40], it is assumed that $\Phi_i = \lambda_i I_{m_i}$, where $0 \leq \lambda_i < 1$ is a given constant value. Owing to the special form of Φ_i, it can commute with S_i and then the control law $u_i(k)$ in (6.13) can be written as

$$u_i(k) = -(S_iB_i)^{-1}S_i[A_{ii}^{\lambda_i}\hat{x}_i(k) + \sum_{j=1,\, j\neq i}^{j=h} \gamma_{ij}A_{ij}\hat{x}_j(k)] - \hat{f}_i(k), \tag{6.14}$$

where $A_{ii}^{\lambda_i} = A_{ii} - \lambda_i I_{n_i}$. Then the compact control law is

$$u(k) = -(SB)^{-1}S(\Gamma \circ A_\lambda)\hat{x}(k) - \hat{f}(k), \tag{6.15}$$

where $A_\lambda = A - \text{diag}[\lambda_i I_{n_i}]_{i=1}^h$. It is worth mentioning that, referring to e.g. [83, 133, 32], the DSMC does not necessarily require a switching component and the linear part in (6.15) leads to a boundary layer with thickness $O(T_s)$. By proper consideration of the sampling phenomenon in the discrete-time sliding mode control design, the boundary layer thickness can be reduced to $O(T_s^2)$ [32]. Moreover, the controller in (6.15) is indeed based on the equivalent control, by removing unknown uncertainty terms and taking into account the structure constraint. The removed terms can be taken care of by robust control techniques.

Remark 6.2 *With different structure matrix Γ, the above controller and the observer in (6.9) can explain various topologies. The decentralized control strategy can be obtained by $\gamma_{ij} = \begin{cases} 1 & \text{if } i = j \\ 0 & \text{otherwise} \end{cases}$ which means that the local controllers use only local state information to control the given subsystem. In this case, there is no control network in the system. When $\Gamma = \mathsf{S}(A)$, each subsystem uses its own state as well as the states of all other physically coupled subsystems. In other words, the control network is structurally the same as the plant network. As the third alternative, the structure matrix Γ can generate a middle-of-the-road solution, between fully distributed control approaches and decentralized ones, regarded as sparsely distributed control systems. This could be beneficial when some constraints on communication requirements between local controllers exist and hence the control network could not have the same structure as the plant network.*

Remark 6.3 *The control network structure will always be a subset of the plant network structure. In other words, if $A_{ij} = 0$ (sub-system j does not influence i-th subsystem), then $\gamma_{ij} = 0$.*

Define the overall state estimation error as

$$e(k) := x(k) - \hat{x}(k), \tag{6.16}$$

and disturbance estimation error as

$$e_f(k) := f(k) - \hat{f}(k). \tag{6.17}$$

The overall closed-loop system is obtained by applying the controller (6.15) to (6.6) and using (6.16), (6.17) and (6.9), as

$$\begin{cases} x(k+1) = (A + \Delta A - \hat{A}_{\Gamma,\lambda})x(k) + B(SB)^{-1}S\left[\Gamma\circ A_\lambda \;\; B\right]e_t(k) \\ e_t(k+1) = \left[\begin{smallmatrix} A+\Delta A-\Gamma\circ A \\ 0 \end{smallmatrix}\right]x(k) + (A_t - L_t C_t)e_t(k) + \bar{f}(k+1), \end{cases} \tag{6.18}$$

where $\hat{A}_{\Gamma,\lambda} = B(SB)^{-1}S(\Gamma \circ A_\lambda)$, $\bar{f}(k+1) = \left[\begin{smallmatrix} 0 \\ \tilde{f}(k+1) \end{smallmatrix}\right]$ with $\tilde{f}(k) = \text{col}(\tilde{f}_i(k))_{i=1}^h$, $e_t(k) = \left[\begin{smallmatrix} e(k) \\ e_f(k) \end{smallmatrix}\right]$, $A_t = \left[\begin{smallmatrix} \Gamma\circ A & B \\ 0 & I_m \end{smallmatrix}\right]$ with $m = \sum_{i=1}^h m_i$, $L_t = \left[\begin{smallmatrix} L_s \\ D_s \end{smallmatrix}\right]$ and $C_t = [C \;\; 0]$.

Lemma 6.3 ([32]) *If the matrix pair $(\Gamma \circ A, C)$ is observable and $(\Gamma \circ A, BC)$ satisfies the rank condition in Assumption 6.2, then the matrix pair (A_ι, C_ι) is observable.*

Also, it can simply be found that

$$\sigma(k+1) = S(\Delta A + A - \Gamma \circ A_\lambda)x(k) + S\left[\Gamma \circ A_\lambda \;\; B\right] e_\iota(k). \tag{6.19}$$

6.4 Stability Analysis

In the case of applying DSMC to the system involving exogenous disturbances, it can only ensure the state trajectories to be driven into a boundary layer around the ideal sliding surface $\sigma(k) = 0$. This issue is indeed regarded as the quasi sliding mode (QSM) in the literature. The following theorem considers a method to analyze simultaneously the reachabiltiy of QSM and the stability of the system states utilizing a discrete-time Lyapunov stability method, in the absence of exogenous disturbances. The characterization of the bounds on the closed-loop system states and sliding function's boundary layer are presented separately later in Theorem 6.2. Furthermore, as Theorem 6.2 needs to derive the cross terms between the system state (sliding function) and the components $\bar{f}(k+1)$, in order to avoid unnecessary repetition of the technical manipulations, we will start the proof of Theorem 6.1 more generally (with the external disturbance and the component $\hat{f}(k)$ in the controller) for the sake of Theorem 6.2. We then let $f(k) = 0$, $\tilde{f}(k) = 0$, and thus, $\bar{f}(k+1) = 0$ to derive the LMI condition for the stability analysis, and control/observer synthesis.

Theorem 6.1 *In the absence of disturbance $f(k)$, the linear part of the control law (6.15) can drive the system state onto the ideal sliding surface (6.10), and the system state is stabilized, if there exist matrices $P = diag[P_i]_{i=1}^h$, with $0 < P_i := U_i^T \begin{bmatrix} P_{i_{11}} & 0 \\ 0 & P_{i_{22}} \end{bmatrix} U_i$, $P_i \in \mathbb{R}^{n_i \times n_i}$, $Q = \begin{bmatrix} diag[\bar{Q}_i]_{i=1}^h & diag[\hat{Q}_i]_{i=1}^h \\ diag[\hat{Q}_i]_{i=1}^{h^T} & diag[\tilde{Q}_i]_{i=1}^h \end{bmatrix} > 0$, with $\bar{Q}_i \in \mathbb{R}^{n_i \times n_i}$, $\tilde{Q}_i \in \mathbb{R}^{m_i \times m_i}$, $\hat{Q}_i \in \mathbb{R}^{n_i \times m_i}$, X_1, X_2 and $X_3 = \begin{bmatrix} \Gamma \circ X_L \\ \Gamma \circ X_D \end{bmatrix}$, with $X_L \in \mathbb{R}^{n \times p}$, $X_D \in \mathbb{R}^{m \times p}$, and scalars $\varepsilon_{ij} > 0$, $i = 1, \cdots, h$, $j = 1, \cdots, h$, and $\rho > 0$ satisfying the following LMI,*

$$\begin{bmatrix}
\check{\chi}_{11} & \star & \star & \star & \star & \star & \star & \cdots & \star \\
0 & \bar{\chi}_{22} & \star & \star & \star & \star & \star & \cdots & \star \\
\sqrt{2}B^T P(A-\Gamma\circ A_\lambda) & \sqrt{2}B^T P\left[\Gamma\circ A_\lambda \;\; B\right] & -B^T PB & \star & \star & \star & \star & \cdots & \star \\
Q\left[\begin{smallmatrix} A-\Gamma\circ A \\ 0 \end{smallmatrix}\right] & QA_t - X_3 C_t & 0 & -Q & \star & \star & \star & \cdots & \star \\
PA+BX_1 & 0 & 0 & 0 & -P & \star & \star & \cdots & \star \\
BX_2 & 0 & 0 & 0 & 0 & -P & \star & \cdots & \star \\
0 & 0 & \sqrt{2}M_1^T PB & \left[M_1^T \;\; 0\right]Q & M_1^T P & 0 & -\Upsilon_1 & \cdots & \star \\
\vdots & \vdots & \vdots & \vdots & \vdots & \vdots & \vdots & \ddots & \vdots \\
0 & 0 & \sqrt{2}M_h^T PB & \left[M_h^T \;\; 0\right]Q & M_h^T P & 0 & 0 & \cdots & -\Upsilon_h
\end{bmatrix} < 0, \tag{6.20}$$

where $0 < P_{i_{11}} \in \mathbb{R}^{m_i \times m_i}$, $0 < P_{i_{22}} \in \mathbb{R}^{(n_i - m_i)\times(n_i - m_i)}$ *and* $U_i \in \mathbb{R}^{n_i \times n_i}$ *is defined in Lemma 3.1,* $\check{\chi}_{11} = -P + X_2^T B^T + BX_2 + \rho I + \sum_{i=1}^{h} \Upsilon_i N_i^T N_i$*, with* $\Upsilon_i = diag[\varepsilon_{ij} I_{n_j}]_{j=1}^{h}$*,* $\bar{\chi}_{22} = -Q + \rho I$*,* $\Gamma = [\gamma_{ij}]_{h\times h}$ *is a given structure matrix and* $\{\star\}$ *denotes the symmetric elements in a symmetric matrix. Here* $S = B^T P$ *and the observer gain is*

$$L_t = Q^{-1} X_3. \tag{6.21}$$

Proof 17 *Define*

$$V(\zeta(k)) = x^T(k)Px(k) + e_t^T(k)Qe_t(k) + \sigma^T(k)(SB)^{-1}\sigma(k), \tag{6.22}$$

where $\zeta(k) = \left[x^T(k) \;\; e_t^T(k) \;\; \sigma^T(k)\right]^T$, $P > 0$ *and* $Q > 0$ *are symmetric matrices and* $S = B^T P$*. Note that the inclusion of both state* $x(k)$ *and sliding function* $\sigma(k)$ *in the Lyapunov candidate function makes it possible to analyze simultaneously the reachabiltiy of QSM as well as the boundedness of the system state and sliding function, as will be seen later in the proof of Theorem 6.2. Thus, it can be written*

$$\begin{aligned}
&\Delta V(\zeta(k)) \triangleq V(\zeta(k+1)) - V(\zeta(k)) \\
=&x^T(k+1)Px(k+1) + e_t^T(k+1)Qe_t(k+1) \\
&+ \sigma^T(k+1)(SB)^{-1}\sigma(k+1) - x^T(k)Px(k) - e_t^T(k)Qe_t(k) \\
&- \sigma^T(k)(SB)^{-1}\sigma(k).
\end{aligned} \tag{6.23}$$

Defining $\varpi = \left[x^T(k) \;\; e_t^T(k) \;\; \bar{f}(k+1)\right]^T$, (6.23) *can be rearranged as:*

$$\Delta V(\zeta(k)) = \varpi^T(k)\left[\chi_{ij}\right]_{3\times 3}\varpi(k), \tag{6.24}$$

where

$$\begin{aligned}
\chi_{11} =&(A+\Delta A)^T P(A+\Delta A)-(A+\Delta A)^T PB(B^T PB)^{-1}B^T P(A+\Delta A)-PB(B^T PB)^{-1}\\
&\times B^T P-P+2(\Delta A+A-\Gamma\circ A_\lambda)^T PB(B^T PB)^{-1}B^T P(\Delta A+A-\Gamma\circ A_\lambda)\\
&+\begin{bmatrix}\Delta A+A-\Gamma\circ A_\lambda\\ 0\end{bmatrix}^T Q\begin{bmatrix}\Delta A+A-\Gamma\circ A_\lambda\\ 0\end{bmatrix},\\
\chi_{12} =&2(\Delta A+A-\Gamma\circ A_\lambda)^T PB(B^T PB)^{-1}B^T P\begin{bmatrix}\Gamma\circ A_\lambda & B\end{bmatrix}\\
&+\begin{bmatrix}\Delta A+A-\Gamma\circ A_\lambda\\ 0\end{bmatrix}^T Q(A_t-L_tC_t),
\end{aligned}$$

$$\begin{aligned}
\chi_{13} =&\begin{bmatrix}\Delta A+A-\Gamma\circ A_\lambda\\ 0\end{bmatrix}^T Q,\\
\chi_{22} =&2\begin{bmatrix}\Gamma\circ A_\lambda & B\end{bmatrix}^T S^T(SB)^{-1}S\begin{bmatrix}\Gamma\circ A_\lambda & B\end{bmatrix}+(A_t-L_tC_t)^T Q(A_t-L_tC_t)-Q,\\
\chi_{23} =&(A_t-L_tC_t)^T Q,\\
\chi_{33} =&Q.
\end{aligned}$$

Now, in order to analyze the system stability let $\bar{f}(k+1)=0$*. The system will be stable if*

$$\Xi := [\chi_{ij}]_{2\times 2} < -\rho I, \tag{6.25}$$

where $\rho>0$ *is a scalar variable. To consider the feasibility of* (6.25)*, by the Schur complement,* (6.25) *is equivalent to*

$$\begin{bmatrix}
\bar{\chi}_{11} & \star & \star & \star\\
0 & \bar{\chi}_{22} & \star & \star\\
\sqrt{2}B^T P(A+\Delta A-\Gamma\circ A) & \sqrt{2}B^T P\begin{bmatrix}\Gamma\circ A_\lambda & B\end{bmatrix} & -B^T PB & \star\\
Q\begin{bmatrix}A+\Delta A-\Gamma\circ A\\ 0\end{bmatrix} & Q(A_t-L_tC_t) & 0 & -Q
\end{bmatrix}<0, \tag{6.26}$$

where

$$\begin{aligned}
\bar{\chi}_{11} =&(A+\Delta A)^T P(A+\Delta A)-(A+\Delta A)^T S^T(SB)^{-1}S(A+\Delta A)\\
&-S^T(SB)^{-1}S-P+\rho I,\\
\bar{\chi}_{22} =&-Q+\rho I.
\end{aligned}$$

Consequently, Lemma 2.3 can be used to show that the feasibility of (6.26) *is equivalent to that of*

$$\begin{bmatrix}\hat{\chi}_{11} & \cdots\\ \vdots & \ddots\end{bmatrix}<0, \tag{6.27}$$

with

$$\begin{aligned}
\hat{\chi}_{11} =&(A+\Delta A+BF_1)^T P(A+\Delta A+BF_1)-P\\
&+F_2^T(B^T PB)F_2+F_2^T B^T P+PBF_2+\rho I,
\end{aligned} \tag{6.28}$$

where F_1 and F_2 are auxiliary variables [86] and note that except $\hat{\chi}_{11}$, other elements of (6.27) *are the same as those in* (6.26). *Therefore, using Lemma 3.1, $\hat{\chi}_{11}$ in* (6.28) *can be rearranged as*

$$
\begin{aligned}
\hat{\chi}_{11} =& [P(A+\Delta A)+BZF_1]^T P^{-1}[P(A+\Delta A)+BZF_1]-P \\
&+F_2^T Z^T B^T P^{-1} BZF_2 + F_2^T Z^T B^T + BZF_2 + \rho I,
\end{aligned}
\tag{6.29}
$$

where Z satisfies $PB = BZ$. Defining $X_1 = ZF_1$, $X_2 = ZF_2$ and $X_3 = QL_t$, with the help of the Schur complement and Lemma 2.1, it can be seen that (6.27) *is sufficed by the LMI in* (6.20).

While the above theorem gives a method to design the structural ODSMC in order to stabilize the overall system in (6.6), it does not present a bound on the system states. The following theorem aims to characterize the boundedness of the obtained overall closed-loop system state and corresponding sliding function in the presence of disturbance $f(k)$.

Theorem 6.2 *In the presence of disturbance $f(k)$, if the LMI in* (6.20) *is feasible, for the obtained P, Q, $L_t = Q^{-1}X_3$ and ρ, the controller* (6.15) *satisfying* (6.7) *will lead to a bound on the augmented system state $\zeta(k) = [x^T(k), e_t(k), \sigma^T(k)]^T$ as follows:*

$$
\forall \varsigma > 0,\ \exists k^\star > 0,\ s.t.\ \forall k > k^\star,\ \ \|\zeta(k)\|^2 \leq \frac{\lambda_{\max}(\mathbf{M})}{\hat{\rho}\lambda_1}\delta + \varsigma,
\tag{6.30}
$$

where $\lambda_1 = \lambda_{\min}(diag(P, Q, (B^T PB)^{-1}))$, $\mathbf{M} = diag(M_P, Q)$, $M_P = PB(B^T PB)^{-1}B^T P + P$, and $\delta = \|\Pi + Q\| \sum_{i=1}^{h} \tilde{L}_i^2 T_s^2$; here the scalar variable $\hat{\rho} > 0$ and matrix variable $\Pi > 0$ are obtained from solving the following LMI:

$$
\begin{bmatrix}
\Omega_1 & \star & \star & \star & \cdots & \star \\
0 & (\hat{\rho}-\rho)I & \star & \star & \cdots & \star \\
\bar{\chi}_{13}^T & \chi_{23}^T & -\Pi & \star & \cdots & \star \\
0 & 0 & \begin{bmatrix} M_1^T & 0 \end{bmatrix} Q & -\bar{\Upsilon}_1 & \cdots & \star \\
\vdots & \vdots & \vdots & \vdots & \ddots & \vdots \\
0 & 0 & \begin{bmatrix} M_h^T & 0 \end{bmatrix} Q & 0 & \cdots & -\bar{\Upsilon}_h
\end{bmatrix} < 0,
\tag{6.31}
$$

where $\bar{\chi}_{13} = \begin{bmatrix} A-\Gamma\circ A_\lambda \\ 0 \end{bmatrix}^T Q$, $\chi_{23} = (A_t - L_t C_t)^T Q$, $\Omega_1 = (\hat{\rho}-\rho)I + \sum_{i=1}^{h} \bar{\Upsilon}_i N_i^T N_i$ and $\bar{\Upsilon}_i = diag[\bar{\varepsilon}_{ij} I_{n_j}]_{j=1}^{h}$ in which $\bar{\varepsilon}_{ij} > 0$, $i = 1,\cdots,h$, $j = 1,\cdots,h$ are scalar variables.

Proof 18 *Defining $\nu(k) = \begin{bmatrix} x^T(k) & e_t^T(k) \end{bmatrix}^T$ and $\chi_v = \begin{bmatrix} \chi_{13}^T & \chi_{23}^T \end{bmatrix}^T$ and according to Lemma 4 in [103] it can be written that*

$$
2\nu^T(k)\chi_v \bar{f}(k+1) \leq \nu^T(k)\chi_v \Pi^{-1} \chi_v^T \nu(k) + \bar{f}^T(k+1)\Pi \bar{f}(k+1),
\tag{6.32}
$$

where $\Pi > 0$ is of appropriate dimension matrix. It follows from (6.24), (6.25) *and*

(6.32) *that*

$$\Delta V(\zeta(k)) \leq -\nu^T(k)[\rho I - \chi_\nu \Pi^{-1} \chi_\nu^T]\nu(k) + \bar{f}^T(k+1)[\chi_{33} + \Pi]\bar{f}(k+1). \tag{6.33}$$

If we choose $\Pi > 0$ *such that*

$$\hat{\rho} I < \rho I - \chi_\nu \Pi^{-1} \chi_\nu^T, \tag{6.34}$$

where $\rho > \hat{\rho} > 0$, *which is always possible if* $\rho > 0$ *exists, then, it follows from* (6.33) *that*

$$\Delta V(\zeta(k)) \leq -\hat{\rho}\nu^T(k)\nu(k) + \bar{f}^T(k+1)[\chi_{33} + \Pi]\bar{f}(k+1). \tag{6.35}$$

Moreover, note that

$$V(\zeta(k)) = \nu^T(k)\mathbf{M}\nu(k), \tag{6.36}$$

where $\mathbf{M} = diag(M_P, Q)$, *and* $M_P = PB(B^T PB)^{-1} B^T P + P$, *hence,*

$$\lambda_{\min}(\mathbf{M}) \|\nu(k)\|^2 \leq V(\zeta(k)) \leq \lambda_{\max}(\mathbf{M}) \|\nu(k)\|^2. \tag{6.37}$$

Additionally, it is known that

$$\lambda_1 \|\zeta(k)\|^2 \leq V(\zeta(k)) \leq \lambda_2 \|\zeta(k)\|^2. \tag{6.38}$$

where $\lambda_1 = \lambda_{\min}(diag(P, Q, (B^T PB)^{-1}))$ *and* $\lambda_2 = \lambda_{\max}(diag\ (P, Q, (B^T PB)^{-1}))$. *Hence, from* (6.35) *and* (6.37) *one can derive that*

$$\Delta V(\zeta(k)) \leq -\frac{\hat{\rho}}{\lambda_{\max}(\mathbf{M})} V(\zeta(k)) + \delta, \tag{6.39}$$

where $\delta = \|\Pi + Q\| \sum_{i=1}^{h} \tilde{L}_i^2 T_s^2$. *Moreover, from* (6.25) *it can simply be written that* $\forall\ \nu(k) \neq 0$

$$\nu^T(k)\Xi\nu(k) = V(\zeta(k+1))\big|_{\bar{f}(k+1)=0} - V(\zeta(k)) < -\rho\nu^T(k)\nu(k). \tag{6.40}$$

It is known that $V(\zeta(k+1))\big|_{\bar{f}(k+1)=0} \geq 0$, *and thus, from* (6.40) *and* (6.37), *it can be claimed that* $\rho < \lambda_{\max}(\mathbf{M})$. *Therefore,* $\frac{\hat{\rho}}{\lambda_{\max}(\mathbf{M})} < 1$. *Finally, from [77, Theorem 5.1, Corollaries 5.1, 5.2] and* (6.39), *one can find the bound given in* (6.30). *Moreover, to find* $\Pi > 0$ *in* (6.34), *for given* $P > 0$, $Q > 0$, $L_i = Q^{-1}X_3$ *and* $\rho > 0$, *by using Lemma 2.1, we can show that* (6.34) *is sufficed by the LMI in* (6.31).

As seen in the proposed sparsely distributed ODSMC, local controllers/observers are able to utilize some interconnections in the nominal A matrix, and the remaining interconnections in A matrix together with ΔA are considered as the uncertainties of the overall system. As an illustration, when $\Gamma = S(A)$ the uncertain term $A + \Delta A - \Gamma \circ A_\lambda$ is $\Delta A + \lambda I_n$, and for the decentralized control network i.e. $\Gamma = I_n$, we have $A + \Delta A - \Gamma \circ A_\lambda = \Delta A + \lambda I_n + A_{off}$, where $A_{off} = A - A_D$, with $A_D = \text{diag}[A_{ii}]_{i=1}^h$; here A_{off} includes all the existing interconnections among the subsystems. The second step of this chapter will consider the issue of minimizing the costs of a control/observer network utilized for the stabilizing distributed ODSMC that can stabilize the underlying large scale system. This will be the subject of the next section.

Remark 6.4 *It is easy to realize from Lemma 6.1 and 6.2 that* $\mathsf{S}(Q^{-1}) = \mathsf{S}(Q)$, *and thus* $\mathsf{S}(L_s) \subseteq \Gamma$, $\mathsf{S}(D_s) \subseteq \Gamma$.

The solution of the LMI in (6.31) does not have direct influence on the controller design and the actual bound on the system state and/or sliding function, however, these parameters may lead us to determine a more accurate bound. Therefore, to obtain the minimum value of the bound in (6.30), the LMIs in (6.20) and (6.31) can be solved subject to a specific criteria. This issue is beyond the scope of this chapter and remains for the future work.

6.5 Sparsifying the Control Network Structure

Previous sections have studied the problem of designing ODSMC for NCSs with imposing *a priori* constraints on communication requirements between sub-systems. In other words, the structure matrix Γ in (6.20) was assumed to be a known one and hence the derived stability condition was an LMI. This section aims to design a control network with a minimum number of links that satisfies the stability condition (6.20). Indeed, the main objective is the minimization of the cost of the control network utilized to stabilize the system. Here, as we will assume that the general cost, including the construction and data transferring costs etc, are identical for all the links, minimizing the costs of a control network can intuitively be considered as minimizing the number of links in the control network structure or equivalently finding the sparsest control network structure that can stabilize the system. We formulate this problem as

$$\min_{P,Q,X_1,X_2,X_3,\rho,\Upsilon_1,\cdots,\Upsilon_h} \quad \textbf{card}(\Gamma) \qquad (6.41)$$
$$\text{subject to (6.20) and } \Gamma \subseteq \mathsf{S}(A),$$

where $\Gamma = [\gamma_{ij}]_{h\times h}$ and $\textbf{card}(\cdot)$ denotes the cardinality function (the number of nonzero elements of a matrix). The above optimization problem is a convex mixed-binary problem which is broadly speaking NP-hard. A number of exact schemes for addressing the convex mixed-binary programs are considered [58]. However, exploiting these schemes is computationally expensive for large networks and in the worst case it may require solving 2^{N} convex problem in order to find the sparsest structure, where N denotes the number of physical interconnections in the plant network. Instead, in this chapter, we will consider a heuristic sub-optimal scheme to deal with this problem.

Notice that the cardinality function, in optimization problems such as (6.41), is usually approximated by the ℓ_1 norm of the optimization variable [30] or the so-called weighted ℓ_1 norm [31]. Since the weighted ℓ_1 norm is not implementable (the required weights should be calculated based on the unknown feedback gain), a reweighted algorithm has recently been proposed in [31], and further used by [88] to design sparse feedback gains. This algorithm solves weighted optimization problems iteratively in which the weights are updated inversely proportional to the strength of individual (block) entries of feedback gain in the previous iteration. Two main reasons restrict the application of the reweighted algorithm to our problem. First, since Γ, in (6.41), is a binary matrix variable, the existing reweighted algorithms are not applicable to this problem. Next, one may also suggest to deal with this problem by defining $K = -(SB)^{-1}S(\Gamma \circ A_\lambda)$ and employing the reweighted algorithm to identify sparse patterns for K. However, this also suffers from the drawback that every obtained solution may not be a feasible solution to the sliding mode controller. This is because since the set of the closed-loop poles may include purely complex conjugate pairs, it cannot be split between the null-space (sliding mode) and range-space dynamics [40]. Moreover, as the problem in this chapter is to sparsify the controller/observer network, updating the weights only according to the off-diagonal block entries of K (or observer gains L_s or D_s) may not result in an adequate solution.

Alternatively, a systematic way of removing the links can be considered by relaxing the constraint on $\gamma_{ij}, i \neq j$ from binary variables to the constraint $0 \leq \gamma_{ij} \leq 1$. We can address the minimization problem (6.41) with the following algorithm:

Algorithm 6.1

1) *Set* $\Gamma = I$. *If the LMI* (6.20) *is feasible,* $\{\gamma^\star_{ij}\} \leftarrow \{\gamma_{ij}\}$, *no control network is required and the sparsest structure is the decentralized structure. Terminate the search and go to Step 6.*

2) *Initialize* $\Gamma = \mathsf{S}(A)$ *and* $l = 1$, *in which* l *denotes the iteration number.*

3) *Solve the LMI* (6.20). *If it is feasible,* $\{\gamma^\star_{ij}\} \leftarrow \{\gamma_{ij}\}$. *Otherwise, if* $l = 1$ *terminate the search and the problem has no solution, or else go to Step 6.*

4) *With known* $P, Q, X_1, X_2, X_3, \rho, \Upsilon_1, \cdots, \Upsilon_h$ *and replacing those entries* $\gamma_{ij} = 1,\ i \neq j$ *with the relaxed constraint* $0 \leq \gamma_{ij} \leq 1$, *minimize* $\sum_{i,j=1,\ i\neq j}^{h} \gamma_{ij}$ *subject to the LMI* (6.20) *and* $0 \leq \gamma_{ij} \leq 1$ *to find* γ^r_{ij}. *Sort the set* $\{\gamma^r_{ij}\}$ *in ascending order.*

5) *Set* γ_{ij} *corresponding to the* l*-th entry of* $\{\gamma^r_{ij}\}$ *to zero and* $l = l + 1$. *If* $l < \mathbf{Card}(\mathsf{S}(A)) - n$ *return to Step 3, otherwise go to Step 6.*

6) *Return* $\gamma^\star_{ij}$.

In the above algorithm, Step 4 characterizes the contribution of each link in the stability of the overall system. Moreover, as seen, the algorithm searches for the sparsest structure using the sorted set $\{\gamma^r_{ij}\}$. In the worst case, in order to find the solution, $2\,(\mathbf{Card}(\mathsf{S}(A)) - n)$ convex problems may be addressed. Finally, it should be stressed that this alternate scheme is only a sub-optimal method to deal with the sparsification problem considered in this section. Broadly speaking, to obtain the optimal solution, one should solve the original mixed-binary convex problem in (6.41), which is NP-hard.

Remark 6.5 *Notice that the minimization algorithm given above does not consider constraints on the control effort level that each sub-system can practically afford. Indeed, this is a common drawback of the current SMC design schemes, the approach of this chapter included, in that the control action required to induce and maintain sliding cannot be taken into account. While the paper [40] considers this problem and proposes an LMI method which can design the sliding function whereby quantifying the control effort, the proposed scheme in [40] is not applicable to our problem which is seeking the sparsest distributed DSMC. This is because the coordinates change proposed in [40] may not give beneficial results. In the next chapter, we will develop a new procedure to solve the Algorithm 6.1 while quantifying the control action.*

6.6 Numerical Examples

6.6.1 Example 1

Consider an interconnected system consisting of three inverted pendulums that are mounted on coupled carts [114, 104], shown in Fig. 6.1. The linearized equations of motions are [114]:

$$\begin{aligned} M_i l\ddot{\theta}_i &= (M_i + m)g\theta_i + c_i\dot{\mathbf{x}}_i + \sum_{j=1,\ j\neq i}^{h} [b_{ij}(\dot{\mathbf{x}}_i - \dot{\mathbf{x}}_j) + k_{ij}(\mathbf{x}_i - \mathbf{x}_j)] - u_i, \\ M_i\ddot{\mathbf{x}}_i &= -c_i\dot{\mathbf{x}}_i - \sum_{j=1,\ j\neq i}^{h} [b_{ij}(\dot{\mathbf{x}}_i - \dot{\mathbf{x}}_j) + k_{ij}(\mathbf{x}_i - \mathbf{x}_j)] - mg\theta_i + u_i, \end{aligned} \tag{6.42}$$

where $k_{ij} = k_{ji}$, $b_{ij} = b_{ji}$, c_i and l are spring, damper, friction coefficients, and pendulum length, respectively. Here, it is assumed that the moment of inertia of the pendulums is zero. Define $x_i = [x_{i,1}, x_{i,2}, x_{i,3}, x_{i,4}]^T = [\theta_i, \dot{\theta}_i, \mathbf{x}_i, \dot{\mathbf{x}}_i]^T$, and now the system in (6.42) is rearranged as an interconnected state space in continuous-time with

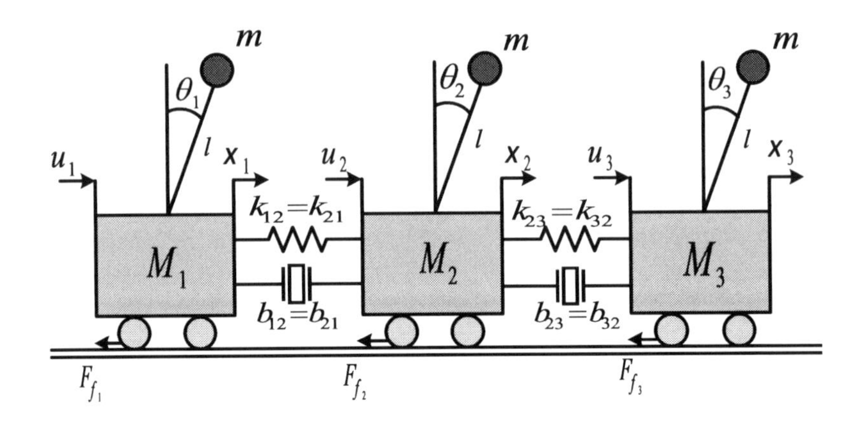

FIGURE 6.1
Three coupled inverted pendulums system.

the system matrices as below:

$$A_{ii} = \begin{bmatrix} 0 & 1 & 0 & 0 \\ \frac{M_i+m}{M_i l}g & 0 & \frac{k_i}{M_i l} & \frac{c_i+b_i}{M_i l} \\ 0 & 0 & 0 & 1 \\ \frac{-m}{M_i}g & 0 & \frac{-k_i}{M_i} & \frac{-c_i-b_i}{M_i l} \end{bmatrix}, \; A_{ij} = \begin{bmatrix} 0 & 0 & 0 & 0 \\ 0 & 0 & \frac{-k_{ij}}{M_i l} & \frac{-b_{ij}}{M_i l} \\ 0 & 0 & 0 & 0 \\ 0 & 0 & \frac{k_{ij}}{M_i} & \frac{b_{ij}}{M_i} \end{bmatrix}$$

$$B_i = \begin{bmatrix} 0 & \frac{-1}{M_i l} & 0 & \frac{1}{M_i} \end{bmatrix}^T, \; C_i = \begin{bmatrix} 1 & 0 & 0 & 0 \\ 0 & 0 & 1 & 0 \end{bmatrix},$$

for $(i,j) \in \mathbf{E} \triangleq \{(1,2),(2,1),(2,3),(3,2)\}$. Here $k_i = \sum_{j \in \mathsf{J}_i} k_{ij}$ and $b_i = \sum_{j \in \mathsf{J}_i} b_{ij}$, where $\mathsf{J}_i := \{j \mid (i,j) \in \mathbf{E}\}$. The numerical system parameters are assumed as $M_1 = 2$, $M_2 = 1$, $M_3 = 3$, $m = 0.5$, $g = 10$, $l = 0.5$, $k_{12} = k_{21} = 5$, $k_{23} = k_{32} = 15$, $b_{12} = b_{21} = 1$, $b_{23} = b_{32} = 5$, $c_1 = 4$, $c_2 = 2$ and $c_3 = 1$. A discretized representation based on a sample interval of 0.005 s is obtained. We set $\lambda_i = 0.7$. In order to check the robustness properties of the controller, the following uncertainty parameters are considered: $\Xi_{ij} = 0.01 \times \mathbf{1}_{4\times 1}$, $\Lambda_{ij} = -0.01 \times \mathbf{1}_{1\times 4}$, $i,j = 1,2,3$. Algorithm 6.1 is solved and then it is found that the sparsest structure that can satisfy the rank condition in Assumption 6.2 and, more importantly, the stability condition in the LMI (6.20) is the decentralized structure. For comparison, we then exploit an exhaustive search on the binary variables, followed by convex optimization of other variables, and the obtained structure is also the decentralized one. We can see that the proposed sub-optimal algorithm leads to the same result as the optimal solution.

6.6.2 Example 2

Consider the system (6.1) with the following parameters:

$$A=\left[\begin{array}{ccc|cc|cc} 0 & 0.2 & 0 & 0.2 & -0.1 & 0.2 & 0 \\ -0.3 & 1.45 & 0.3 & -0.1 & 0 & 0.03 & 0 \\ 0.3 & 0 & 0.4 & -0.2 & 0.2 & -0.1 & 0 \\ \hline -0.03 & 0.1 & 0.1 & 0.05 & 0.2 & 0 & 0.2 \\ 0 & 0.1 & 0.05 & 0 & 1.3 & 0.1 & -0.1 \\ \hline 0 & 0.2 & 0.1 & 0 & 0 & 0.1 & -0.3 \\ 0 & -0.2 & 0.2 & -0.2 & 0 & 0 & 1.1 \end{array}\right],$$

$$B=\left[\begin{array}{cc|c|c} 1 & 0 & 0 & 0 \\ 0 & 0.5 & 0 & 0 \\ 2 & 0 & 0 & 0 \\ \hline 0 & 0 & 1 & 0 \\ 0 & 0 & 1 & 0 \\ \hline 0 & 0 & 0 & 1 \\ 0 & 0 & 0 & 2 \end{array}\right],\ C=\left[\begin{array}{ccc|cc|cc} -10 & 2 & 1 & 0 & 0 & 0 & 0 \\ 1 & -3 & -1 & 0 & 0 & 0 & 0 \\ \hline 0 & 0 & 0 & 1 & 0 & 0 & 0 \\ 0 & 0 & 0 & -2 & 1 & 0 & 0 \\ \hline 0 & 0 & 0 & 0 & 0 & 2 & 1 \end{array}\right],$$

$$\Xi_{11}=\begin{bmatrix}0.1\\0.02\\0.1\end{bmatrix},\ \Xi_{12}=\begin{bmatrix}0.1\\-0.1\\0\end{bmatrix},\ \Xi_{13}=\begin{bmatrix}0.01\\0\\-0.01\end{bmatrix},\ \Xi_{21}=\begin{bmatrix}0.1\\0.2\end{bmatrix},\ \Xi_{22}=\begin{bmatrix}0\\-0.2\end{bmatrix},$$

$$\Xi_{23}=\begin{bmatrix}-0.01\\0\end{bmatrix},\ \Xi_{31}=\begin{bmatrix}0.2\\-0.01\end{bmatrix},\ \Xi_{32}=\begin{bmatrix}-0.01\\0\end{bmatrix},\ \Xi_{33}=\begin{bmatrix}0.1\\0.2\end{bmatrix},$$

$$\Lambda_{11}^T=\begin{bmatrix}0.1\\0.1\\-0.1\end{bmatrix},\ \Lambda_{21}^T=\begin{bmatrix}0.08\\0.1\\0.1\end{bmatrix},\ \Lambda_{31}^T=\begin{bmatrix}0.01\\-0.02\\0\end{bmatrix},\ \Lambda_{12}^T=\begin{bmatrix}0.05\\0\end{bmatrix},\ \Lambda_{22}^T=\begin{bmatrix}0.05\\0.2\end{bmatrix},$$

$$\Lambda_{32}^T=\begin{bmatrix}0\\0.03\end{bmatrix},\ \Lambda_{13}^T=\begin{bmatrix}0.01\\0.2\end{bmatrix},\ \Lambda_{23}^T=\begin{bmatrix}-0.1\\0.2\end{bmatrix},\ \Lambda_{33}^T=\begin{bmatrix}-0.02\\0\end{bmatrix},$$

$\Theta_{ij}(k)=0.5\sin(k)$, $\lambda=0.8$.

All three open-loop subsystems are unstable. Suppose

$$\xi(k)=\left[0.1\sin(\tfrac{k}{4})\cos(\tfrac{k}{12}),\quad 0.07\sin(\tfrac{k}{5}),\quad 0.1\cos(\tfrac{k}{5})\sin^2(\tfrac{k}{9}),\quad 0.05\sin(\tfrac{k}{8})\cos^2(\tfrac{k}{9})\right]^T.$$

Solving the LMI feasibility problem in (6.20), by assuming a fully distributed structure for the control network; i.e. $\Gamma=\mathrm{S}(A)$, we obtain

$$S=\left[\begin{array}{ccc|cc|cc} 0.0503 & -0.0331 & 0.1005 & 0 & 0 & 0 & 0 \\ -0.0033 & 0.0276 & -0.0066 & 0 & 0 & 0 & 0 \\ \hline 0 & 0 & 0 & 0.0466 & 0.0466 & 0 & 0 \\ \hline 0 & 0 & 0 & 0 & 0 & 0.0768 & 0.0829 \end{array}\right],$$

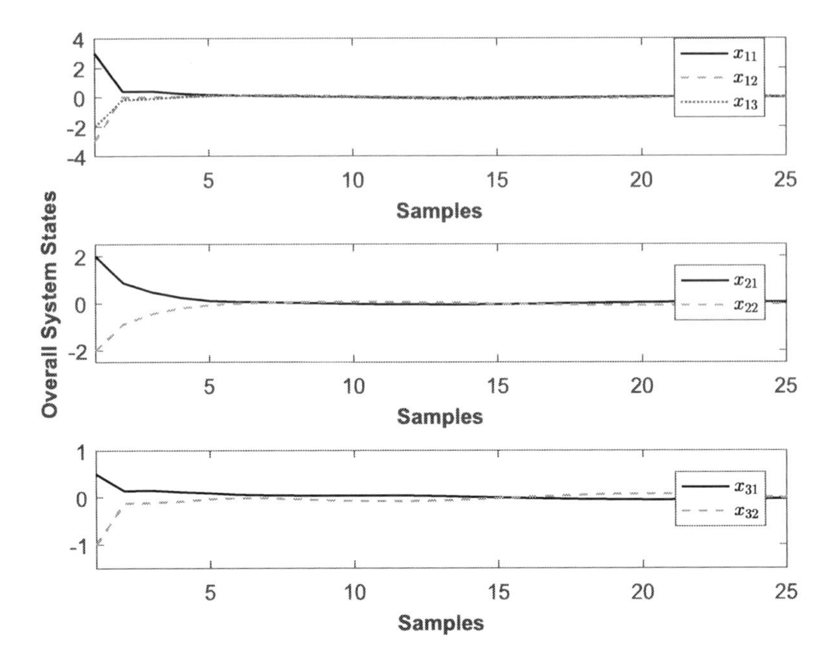

FIGURE 6.2
Trajectories of the system state with fully distributed structure.

$$L = \left[\begin{array}{cc|cc|c} -0.0991 & -0.1359 & 0.0832 & -0.0411 & 0.0466 \\ -0.1752 & -0.7749 & -0.1106 & 0.0314 & 0.0526 \\ -0.2725 & -0.2341 & 0.3745 & 0.3325 & -0.0075 \\ \hline -0.0331 & -0.0892 & 1.3070 & 0.5230 & 0.1154 \\ -0.0342 & -0.1026 & 3.8072 & 1.8234 & -0.0135 \\ \hline -0.0496 & -0.1109 & -0.1630 & -0.0545 & 0.0914 \\ -0.0312 & 0.0566 & -0.5126 & -0.1075 & 0.9784 \end{array}\right],$$

$$D = \left[\begin{array}{cc|cc|c} -0.0759 & -0.0484 & 0.0826 & 0.0537 & -0.0040 \\ 0.0009 & -0.1370 & -0.0544 & -0.0256 & -0.0021 \\ \hline -0.0060 & -0.0172 & 0.7952 & 0.2834 & 0.0039 \\ \hline -0.0083 & 0.0043 & -0.1712 & -0.0637 & 0.2050 \end{array}\right],$$

$$\begin{aligned} \Upsilon_1 &= \text{diag}(0.3183, 0.9880, 0.0793), \\ \Upsilon_2 &= \text{diag}(0.2577, 0.0889, 0.0813), \\ \Upsilon_3 &= \text{diag}(1.6401, 0.5180, 1.1030), \\ \rho &= 4.4631 \times 10^{-4}. \end{aligned}$$

The control law and observer in (6.15) and (6.9) are derived by the achieved S, L, D, $\Gamma^{\star}$, and with $\lambda = 0.8$. Applying this controller to the system (6.1), we illustrate the results in Figs. 6.2 - 6.4. Here, the initial state is assumed to be $x(0) = \begin{bmatrix} 3 & -3 & -2 & 2 & -2 & 0.5 & -1 \end{bmatrix}^T$. Fig. 6.5 shows the performance of the

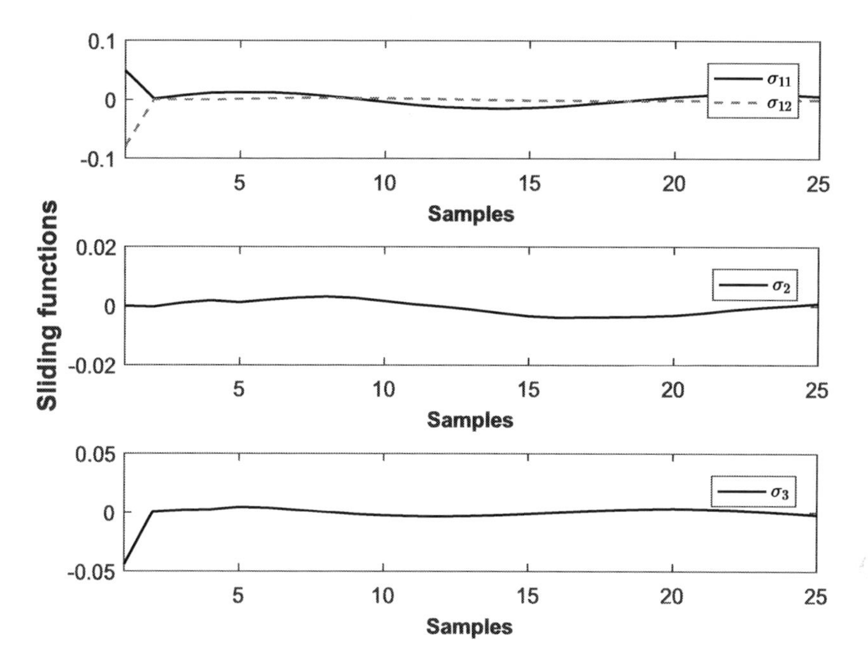

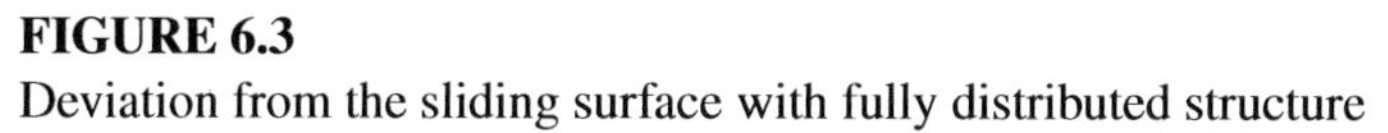

FIGURE 6.3
Deviation from the sliding surface with fully distributed structure

designed disturbance estimator. Now, we consider the problem of identifying the sparsest stabilizing control network structure. Firstly, the LMI in (6.20) is not feasible with the decentralized structure. Following the procedure given in Algorithm 6.1, the sparse structure matrix is obtained as

$$\Gamma^\star = [\gamma^\star_{ij}]_{3\times3} = \begin{bmatrix} 1 & 1 & 0 \\ 0 & 1 & 0 \\ 1 & 1 & 1 \end{bmatrix}.$$

It is worth mentioning that using an exhaustive search instead of the proposed simplified Algorithm 6.1 would result in the same structure. This demonstrates the effectiveness of the proposed sub-optimal algorithm in the chapter for identifying a favorable sparse stabilizing topology for the control network. Solving the LMI feasibility problem in (6.20), with the structure constraint obtained previously ($\Gamma^\star$), gives

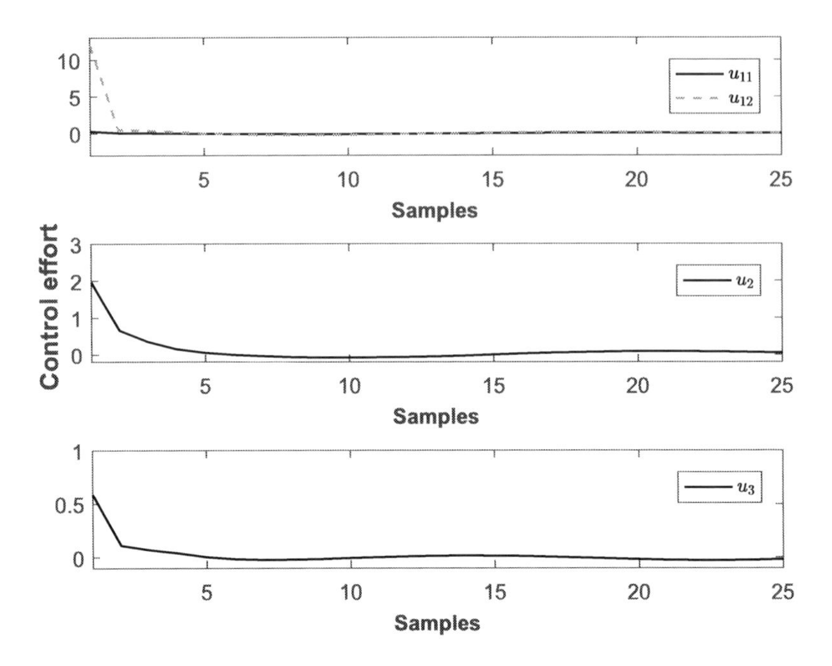

FIGURE 6.4
Control efforts with fully distributed structure.

the parameters below:

$$S = 10^{-4}\times\left[\begin{array}{ccc|cc|cc} 0.0340 & 0.0143 & 0.0679 & 0 & 0 & 0 & 0 \\ 0.0014 & 0.0884 & 0.0029 & 0 & 0 & 0 & 0 \\ \hline 0 & 0 & 0 & 0.0220 & 0.0220 & 0 & 0 \\ \hline 0 & 0 & 0 & 0 & 0 & 0.6066 & 0.6702 \end{array}\right],$$

$$L = \left[\begin{array}{cc|cc|c} 0.0080 & -0.0891 & -0.0994 & -0.0831 & 0 \\ 0.0227 & -0.5051 & -0.1275 & 0.0042 & 0 \\ -0.0149 & -0.0253 & 0.2479 & 0.1975 & 0 \\ \hline 0 & 0 & 0.5244 & 0.4810 & 0 \\ 0 & 0 & 3.0021 & 1.7892 & 0 \\ \hline -0.0069 & -0.0994 & 0.1756 & 0.0148 & -0.2955 \\ 0.0408 & 0.1892 & -0.3746 & -0.0298 & 0.4391 \end{array}\right],$$

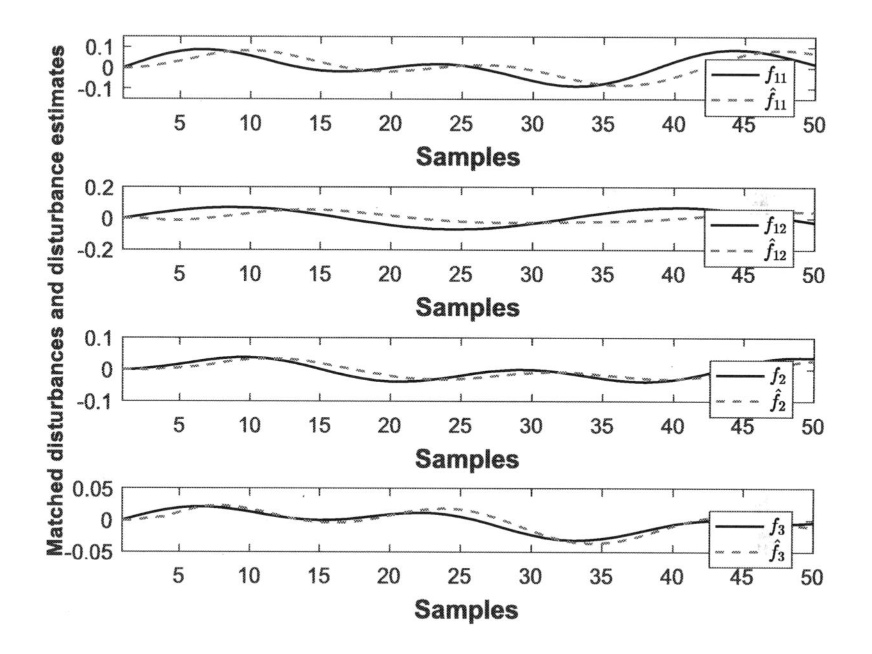

FIGURE 6.5
Exogenous disturbances and disturbance estimator outputs with fully distributed structure.

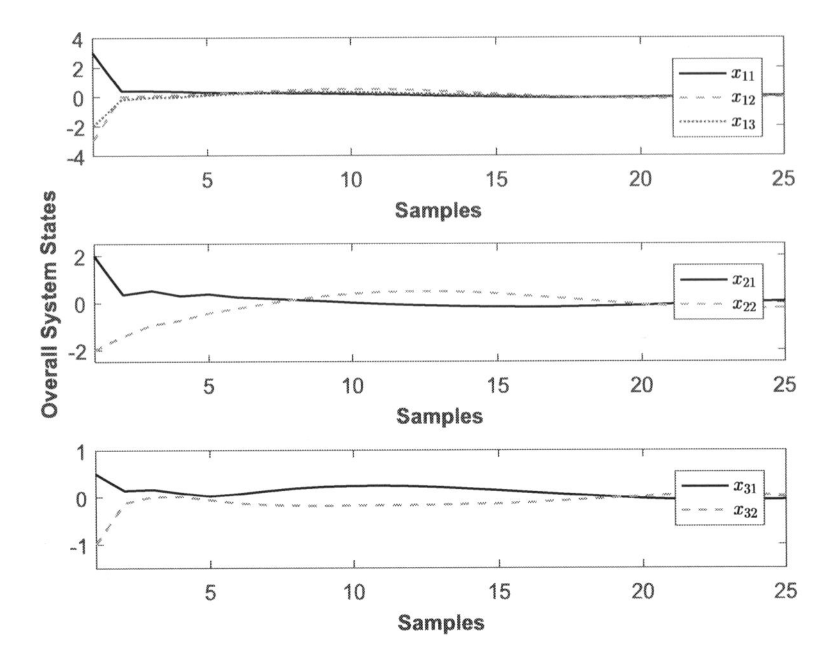

FIGURE 6.6
Trajectories of the system state with structure $\Gamma^{\star}$.

$$
D = \left[\begin{array}{cc|cc|c}
-0.0000 & -0.0020 & -0.0001 & 0.0000 & 0 \\
0.0000 & -0.0006 & 0.0000 & 0.0000 & 0 \\
\hline
0 & 0 & 0.0001 & 0.0001 & 0.00 \\
\hline
0.0000 & 0.0001 & -0.0002 & -0.0000 & 0.0002
\end{array}\right],
$$

$$
\begin{aligned}
\Upsilon_1 =& 10^{-3} \times \text{diag}(0.0390, 0.1996, 0.0079), \\
\Upsilon_2 =& 10^{-5} \times \text{diag}(0.6881, 0.3360, 0.0392), \\
\Upsilon_3 =& \text{diag}(0.0019, 0.0000, 0.0003), \\
\rho =& 4.7193 \times 10^{-11}.
\end{aligned}
$$

Using the given S, L, D, $\Gamma^{\star}$ and $\lambda = 0.8$, the control law and observer given in (6.15) and (6.9), respectively, can be obtained. Figs. 6.6 - 6.8 demonstrate the closed-loop system state trajectories, deviation from sliding surface as well as control efforts, obtained by applying the controller to the system (6.1). As seen from these figures, the proposed sparse ODSMC law can successfully steer the state trajectories into a boundary layer about the ideal sliding surface and keep them there thereafter. However, it should also be noted that with sparser structures, the ultimate bounds on the overall system state trajectories are wider compared to the less sparse structures. To evaluate this issue in a more quantitative way, we define

$$
E_{rms} \triangleq \max_{i}(e_{rms,i}), \quad i = 1, \cdots, m, \tag{6.43}
$$

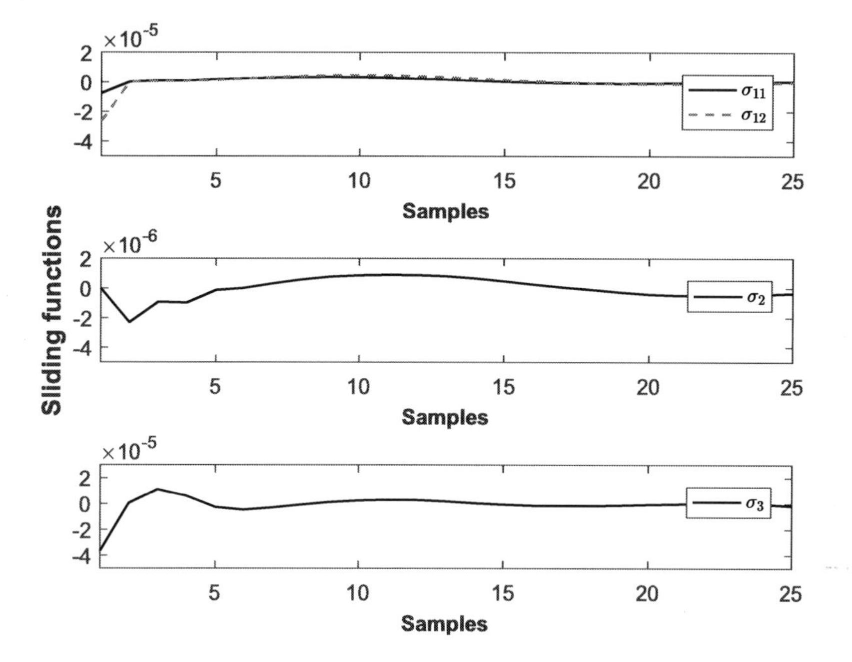

FIGURE 6.7
Deviation from the sliding surface with structure $\Gamma^{\star}$.

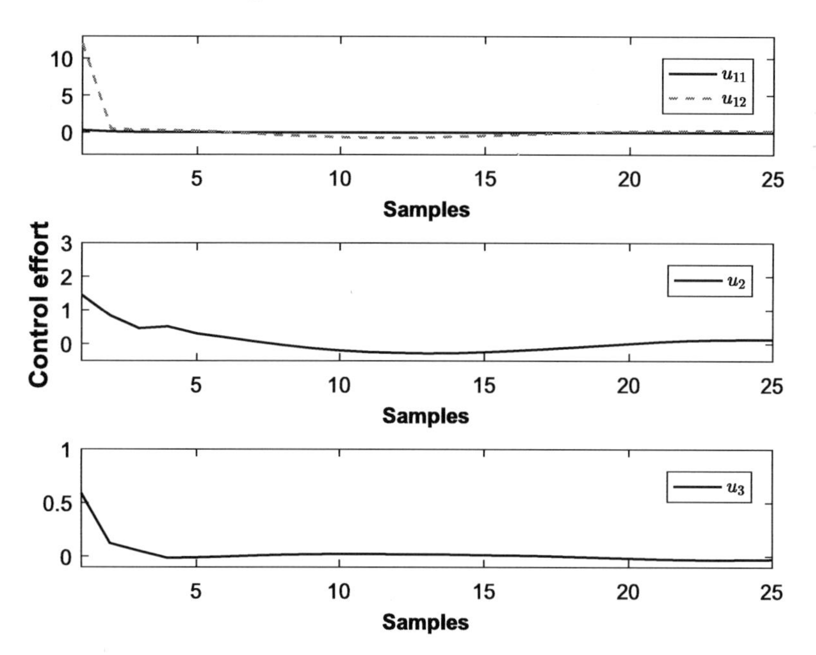

FIGURE 6.8
Control efforts with structure $\Gamma^{\star}$.

where $e_{rms,i}$ is the discrete-time cumulative root mean square (RMS) value of $e_{\xi,i}$, as

$$e_{rms,i} \triangleq \sqrt{\frac{1}{N}\sum_{k=1}^{N} |e_{\xi,i}(k)|^2}, \quad i = 1, \cdots, m, \tag{6.44}$$

where N denotes the number of samples and e_ξ is defined in (6.17). Table 6.1 presents E_{rms} corresponding to different structures. As seen from these results, the less sparse the control network structure, the narrower the ultimate bound on the state trajectories. This can also lead us to this conclusion that a trade off between the control performance and sparsity of the control network should be considered.

TABLE 6.1
Comparison of E_{rms} for different structures

Γ	$\begin{bmatrix} 1 & 1 & 1 \\ 1 & 1 & 1 \\ 1 & 1 & 1 \end{bmatrix}$	$\begin{bmatrix} 1 & 1 & 1 \\ 1 & 1 & 0 \\ 1 & 1 & 1 \end{bmatrix}$	$\begin{bmatrix} 1 & 1 & 0 \\ 1 & 1 & 0 \\ 1 & 1 & 1 \end{bmatrix}$	$\begin{bmatrix} 1 & 1 & 0 \\ 0 & 1 & 0 \\ 1 & 1 & 1 \end{bmatrix}$
E_{rms}	0.0066	0.0068	0.0076	0.0076

6.6.3 Example 3

Consider the system given in Example 2 of [114]. A discretized representation based on a sample interval of 0.01 s is

$$A = \begin{bmatrix} 1.0100 & 0.0090 & 0 \\ 0.0040 & 1.0100 & 0.0060 \\ 0 & 0.0070 & 1.0100 \end{bmatrix}, \quad B = \begin{bmatrix} 0.01 & 0 & 0 \\ 0 & 0.01 & 0 \\ 0 & 0 & 0.01 \end{bmatrix}.$$

We assume $\lambda = 0.5$. To test the robustness properties of our controller, the following uncertainty parameters are considered, $\Xi_{11} = 0.1$, $\Xi_{22} = 0.1$, $\Xi_{33} = 0.1, \Lambda_{11} = 0.1$, $\Lambda_{22} = 0.2$, $\Lambda_{33} = 0.1$, $\Xi_{12} = 0.01$, $\Xi_{13} = .05$, $\Xi_{23} = 0.02$, $\Xi_{21} = .01$, $\Xi_{31} = .01$, $\Xi_{32} = 0.1$, $\Lambda_{12} = 0.01$, $\Lambda_{13} = 0.01$, $\Lambda_{23} = 0.05$, $\Lambda_{21} = 0.02, \Lambda_{31} = 0.05$ and $\Lambda_{32} = 0.05$.

Algorithm 6.1 then suggests a decentralized structure for the control network, which is consistent with the result in [114]. The simulation shows that the controller obtained from solving the LMI in (6.20) by the decentralized structure for Γ can effectively stabilize the overall system.

6.7 Conclusions

This chapter first, with assuming a priori known structure for the control/observer network, proposes a sparse DSMC by utilizing only control system sensors' signals for the networked systems. A unified framework is derived for the observer-based controller design, with the aid of an LMI scheme. Furthermore, our sparse ODSMC reduces the conservatism of the existing methods in the literature for the LMI based DSMC. Then, this chapter explores the solution to the problem of finding the sparsest control/observer network structure that satisfies the LMI stability condition obtained in the first part. A numerical example has been presented to show the effectiveness of the proposed scheme.

7

$\mathscr{H}_2$-Based Optimal Sparse Sliding Mode Control for Networked Control Systems

CONTENTS

7.1 Introduction 141
7.2 Problem Formulation and Preliminaries 143
7.3 Optimal Structured SMC Design Problem 147
7.3.1 $\mathscr{H}_2$ based optimal structured static output feedback 147
7.3.2 Stability analysis of sliding mode dynamics 149
7.4 Sparsification of the Control Network 150
7.5 Numerical Examples 152
7.5.1 Example 1 152
7.5.2 Example 2 153
7.5.2.1 Comparison 1 153
7.5.2.2 Comparison 2 154
7.6 Solving *LQ* SOF Problem 155
7.7 Reweighted ℓ_1 Minimization Algorithm 157
7.8 Conclusions 157

Abstract– This chapter considers the problem of designing a sparsely distributed sliding mode control for networked systems. A distributed sliding mode control framework by exploiting (some of) other subsystems' information is designed to improve the performance of each local controller so that it can widen the applicability region of the given scheme. First, a novel approach is proposed to design the sliding surface, in which the level of required control effort is taken into account during the sliding surface design based on the $\mathscr{H}_2$ control. Then this novel scheme is used to provide an innovative less-complex procedure that explores sparse control networks to satisfy the underlying control objective. Besides, the proposed scheme to design the sliding surface is able to avoid unbounded growth of control effort during the sparsification of the control network structure.

7.1 Introduction

Chapter 6 considers the problem of designing a sparsely distributed observer-based DSMC strategy. First, by assuming a priori topology for the control network, a stabilizing DSMC is developed. Then, a procedure is proposed for seeking a favorable sparse control network structure for the stabilizing DSMC. In other words, in Chapter 6, a sparse control/observer network structure with the least possible number of links has been sought that can satisfy the given stability condition. On the other hand, as it was mentioned in Remark 6.5, although the SMC is now a well-known strategy, from the standpoint of constraining the available control action, all the traditional methods considered in the literature, including the DSMC in Chapter 6, have shortcomings [40]. This drawback basically comes from the nature of the SMC design process which contains two separate stages. During the synthesizing of the sliding function, there is no sense of the control action level that is required to induce and retain sliding. This issue is more crucial in this chapter when it comes to sparsifying the control network structure, as with no limits on the available control actions, it may result in the high level of control efforts that each subsystem's controller is required to apply, which is not a practical case.

To deal with this problem, [105] proposes a scheme to design a sliding surface which minimizes a cost functional of the system state and control input. However, the method given in [105] has several limitations. As the method in [105] needs to ensure that at least one eigenvalue of the closed-loop system (for single-input systems) is a real value, not necessarily any arbitrary weighting matrices in the objective function can result in a sliding mode control. Hence, this reference either reselects the weighting matrices or approximates the closed-loop system eigenvalues so that a set of eigenvalues is generated which can be split between the range-space and null-space dynamics. However, no precise scheme is given on how to reselect the weighting matrices. Furthermore, the approximation of eigenvalues may lead to a loss in optimality and possibly robustness. In order to resolve the limitations of [105], [135] proposes a framework in which a weighting matrix is computed that is tried to be the closest to the desired one and also results in the desired eigenvalues. The SMC then can be designed according to the obtained eigenvalues and weighting matrix. However, both methods in [105, 135] are only applicable to single-input systems. Alternatively, [40] considers this problem and proposes two new frameworks. However, the proposed methods in [40] rely on a special system coordinate transformation which bounds the possible adoption of these methods to our control structure sparsification problem. This chapter alternatively develops a different approach by which we can deal with an $\mathscr{H}_2$ based optimal structured SMC problem.

Recently, the issue of designing a control network with minimum communication links has been studied in the literature [121, 122, 120, 88, 114]. As an illustration, [88] proposes a non-convex condition which is solved numerically by exploiting a convex reweighted ℓ_1 norm approximation. Furthermore, [114] considers the problem of finding the sparsest control/observer network that satisfies the obtained stabil

ity condition. Roughly speaking, the sparsity is formulated in terms of the cardinality (ℓ_0 quasi-norm) of the feedback gain in these references, which is then relaxed by the (weighted) ℓ_1-norm (see [31]). In this chapter in order to address the problem of designing a sparse SMC controller, a specific form of fictitious system, whose matrices contain the control network structure, is derived. This makes the well-developed weighted ℓ_1 algorithm infeasible to our problem. Alternatively, this chapter proposes a heuristic scheme to obtain the sparse sliding mode controller.
Finally, it is worth noting that unlike previous chapters we considered a continuous-time networked system and proposed an optimal sparse continuous-time SMC. However, the results derived in this chapter can simply be extended to discrete-time systems.

The rest of this chapter is as follows. Section 7.2 describes the problem formulation and preliminaries. Section 7.3 presents the structured $\mathscr{H}_2$ based sliding mode control. Section 7.4 demonstrates our heuristic method to solve the problem of finding a favorable sparse structure for the control network. The effectiveness of the proposed sparse distributed SMC is studied by numerical examples in Section 7.5. An algorithm for solving the derived LQ static output feedback problem in Section 7.3 is given in Section 7.6. Section 7.7 includes an ℓ_1 minimization algorithm used for sparsification of feedback gains in Section 7.5. Finally, Section 7.8 concludes this chapter.

7.2 Problem Formulation and Preliminaries

Consider a large scale networked system consisting of h subsystems,

$$\dot{x}_i(t) = A_i x_i(t) + \sum_{j=1,\ j\neq i}^{h} A_{ij} x_j(t) + B_i[u_i(t) + f_i(x_i)], \tag{7.1}$$

where $x_i \in \mathbb{R}^{n_i}$, $u_i \in \mathbb{R}^{m_i}$ and $z_i(t) \in \mathbb{R}^{q_i}$ are the state vector, control input vector and performance output vector of the i-th subsystem, respectively. The matrices in (7.1) are constant and of appropriate dimensions. Besides, $A_{ij} \neq 0$ if the sub-system j influences directly the sub-system i. Without loss of generality, it is also assumed that $\operatorname{rank}(B_i) = m_i$. $f_i(x_i) \in \mathbb{R}^{m_i}$ is the matched uncertainty.

To design the sliding surface in this chapter it is assumed that the system in (2.1) is in a special coordinate (see e.g. [40]) which is more stringent than what is considered in the well-known regular form coordinate. Thus, it is assumed that the input distribution matrix in (2.1) has the following form

$$B_i = \begin{bmatrix} 0 \\ I_{m_i} \end{bmatrix}, \tag{7.2}$$

Define

$$\begin{aligned} x(t) &:= \operatorname{col}(x_i(t))_{i=1}^{h}, \quad u(t) := \operatorname{col}(u_i(t))_{i=1}^{h}, \\ f(x) &:= \operatorname{col}(f_i(x_i))_{i=1}^{h}, \end{aligned} \tag{7.3}$$

and

$$A := \mathrm{diag}[A_i]_{i=1}^h + \left[A_{ij}\right]_{h\times h},\ B := \mathrm{diag}[B_i]_{i=1}^h, \tag{7.4}$$

in which $A_{ii} = 0$. Using (7.1), (7.3) and (7.4), the overall system can be written as

$$\dot{x}(t) = Ax(t) + B[u(t) + f(x)]. \tag{7.5}$$

It is assumed in this chapter that some additional states from other subsystems are utilized to improve the performance of the control loop. This idea is different from the decentralized controller and will lead to a distributed control structure. Note that the control network may differ from the system network. Our objective here is to design an $\mathscr{H}_2$-based optimal distributed SMC, exploiting feedback from (some of) other subsystems, to stabilize the overall system in (7.5) through a sparse control network. Now, consider the following linear sliding function

$$\sigma(t) = Sx(t), \tag{7.6}$$

where $\sigma(t) := \mathrm{col}(\sigma_i(t))_{i=1}^h$ and the block diagonal matrix $S := \mathrm{diag}[S_i]_{i=1}^h$ will be designed later such that SB is nonsingular.

During the ideal sliding motion the sliding function satisfies:

$$\sigma(t) = 0,\ \ \forall t > t_s, \tag{7.7}$$

where $t_s > 0$ denotes the time that sliding motion starts. Due to the special system coordinate explained before, the overall switching function S matrix may be parameterized as

$$S = \mathrm{diag}[\bar{S}_i]_{i=1}^h \cdot \mathrm{diag}\left[\begin{bmatrix} M_i & I_{m_i} \end{bmatrix}\right]_{i=1}^h, \tag{7.8}$$

where $M_i \in \mathbb{R}^{m_i\times(n_i-m_i)}$ and $\bar{S}_i$ are nonsingular matrices having no influence on the overall reduced-order sliding motion. Now, the controller is assumed to be of the following structure:

$$u_i(t) = -(S_iB_i)^{-1}\left\{(S_iA_i - \Phi_iS_i)x_i(t) + S_i\sum_{j=1}^{h}\gamma_{ij}A_{ij}x_j(t)\right\} + \vartheta_i(t), \tag{7.9}$$

where $\Phi_i \in \mathbb{R}^{m_i\times m_i}$ is a stable matrix, γ_{ij} denotes the ij-th element of the structure matrix Γ of the control network, that is, $\gamma_{ij} = 1$ if $i = j$, or the ij-th link exists in the control network and $\gamma_{ij} = 0$ otherwise, and $\vartheta_i(t) \in \mathbb{R}^{m_i}$ denotes the nonlinear part of the controller.

Assumption 7.1 *There exist known continuous functions $\rho_i(\cdot)$ and $\mu_i(\cdot)$ such that for $i = 1,\cdots,h$:*

1) $\|f_i(x_i)\| \le \rho_i(x_i)$,

2) $\left\|\sum_{j=1,j\neq i}^{h}(1-\gamma_{ij})A_{ij}x_j\right\| \le \mu_i(\bar{\gamma}_i^T \circ x)$,

where $\bar{\gamma}_i$ implies the i-th row of $(\mathbb{1}_{h\times h} - \Gamma)$.

Then the nonlinear part of the controller has the following form

$$\vartheta_i(t) = -(S_iB_i)^{-1}\{\|S_iB_i\|\,\rho_i(x_i) + \kappa_i(x_i)\}\frac{\sigma_i(t)}{\|\sigma_i(t)\|} \text{ if } \sigma_i(t) \neq 0, \tag{7.10}$$

in which $\kappa_i(x_i)$ is a gain to be designed later in this section.

Besides, we need to design the sliding function matrix so that the resulting reduced $(n_i - m_i)$ order sliding mode dynamics are stable. Thus, our next problem is to design sliding matrices S_i ensuring overall stability and an additional $\mathscr{H}_2$ performance specification. Notice that the role of the term $(S_iB_i)^{-1}\Phi_iS_ix_i(k)$ in the controller (7.9) is to govern the convergence rate to the sliding manifold in association with the nonlinear part $\vartheta_i(t)$. Here, similar to [40], it is assumed that $\Phi_i = \lambda_iI_{m_i}$, where $\lambda_i < 0$ is a given constant value. Note that unlike in [40], λ_i can also belong to the spectrum of A_i. Owing to the special form of Φ_i, it can commute with S_i and then the control law $u_i(k)$ in (7.9) can be written as

$$u_i(t) = (S_iB_i)^{-1}S_i\left\{A_{\lambda,i}x_i(k) - \sum_{j=1}^{h}\gamma_{ij}A_{ij}x_j(t)\right\} + \vartheta_i(t), \tag{7.11}$$

where $A_{\lambda,i} = A_i - \lambda_iI_{n_i}$. Then the compact control law is

$$u(t) = -(SB)^{-1}S(\Gamma\circ A_\lambda)x(t) + \vartheta(t), \tag{7.12}$$

where $A_\lambda = A - \text{diag}[\lambda_iI_{n_i}]_{i=1}^h$, $\Gamma = [\gamma_{ij}]_{h\times h}$ and $\vartheta(t) = \text{col}(\vartheta_i(t))_{i=1}^h$.

We now aim to show that the controller (7.11), (7.10) drives the system state to the composite sliding surface (7.6). Further in what follows, we assume the known sliding surface matrix $S := \text{diag}[S_i]_{i=1}^h$ and its design will be derived in the next section.

Theorem 7.1 *Consider the NCSs in* (2.1). *Under Assumption 7.1, the sparse controller* (7.11), (7.10) *drives the state of the system* (7.1) *to the composite sliding surface* (7.6) *and maintains a sliding motion if* $\kappa_i(x_i)$ *satisfies*

$$\sum_{i=1}^{h}\kappa_i(x_i) > \sum_{i=1}^{h}\|S_i\|\,\mu_i(\bar{\gamma}_i\circ x) \tag{7.13}$$

where S_i are given sliding function matrices and $\mu_i(\cdot)$ are determined by Assumption 7.1.

Proof 19 *The dynamics of σ_i of subsystem i can be derived by taking the time derivative of* (7.6), *substituting in the state equation* (7.1), *and using the controller* (7.11), (7.10), *i.e.,*

$$\begin{aligned}\dot{\sigma}_i(t) =& \lambda_i\sigma_i(t) + S_i\sum_{j=1,j\neq i}^{h}(1-\gamma_{ij})A_{ij}x_j(t) - [\kappa_i(x_i) + \|S_iB_i\|\,\rho_i(x_i)]\frac{\sigma_i(t)}{\|\sigma_i(t)\|}\\ &+ S_iB_if_i(x_i).\end{aligned} \tag{7.14}$$

Now we will prove that the following composite reachability condition is satisfied [65]:

$$\sum_{i=1}^{h} \frac{\sigma_i^T \dot{\sigma}_i}{\|\sigma_i\|} < 0. \tag{7.15}$$

It follows from (7.14) *and Assumption 7.1 that*

$$\begin{aligned}\frac{\sigma_i^T \dot{\sigma}_i}{\|\sigma_i\|} &\le \lambda_i \|\sigma_i\| + \|S_i\| \, \mu_i(\bar{\gamma}_i \circ x) - \kappa_i(x_i) + \|S_i B_i f_i(x_i)\| - \|S_i B_i\| \, \rho_i(x_i) \\ &\le \|S_i\| \, \mu_i(\bar{\gamma}_i \circ x) - \kappa_i(x_i).\end{aligned} \tag{7.16}$$

Finally, if $\kappa_i(x_i)$ satisfies (7.13), *the composite reachability condition* (7.15) *holds.*

Remark 7.1 *An obvious choice for $\mu_i(\bar{\gamma}_i^T \circ x)$ is $\sum_{j=1, j\neq i}^{h} \left\|(1-\gamma_{ij})A_{ij}\right\| \left\|x_j\right\|$, and as a consequence, $\kappa_i(x_i) = \sum_{j=1, j\neq i}^{h} \left\|S_j\right\| \left\|(1-\gamma_{ji})A_{ji}\right\| \|x_i\| + \varepsilon_i$, with $\varepsilon_i > 0$ a small given scalar, satisfies the condition* (7.13).

Note that thanks to the special structure of S and B, the controller can be written as

$$u(t) = -\text{diag}\left[\begin{bmatrix} M_i & I_{m_i} \end{bmatrix}\right]_{i=1}^{h} (\Gamma \circ A_\lambda) x(t) + \vartheta(t), \tag{7.17}$$

With different structure matrix Γ, the above controller can explain various topologies. The decentralized control strategy can be obtained by $\Gamma = I_h$ which means that the local controllers use only local state information to control the given subsystem. When $\Gamma = \mathsf{S}(A)$ we may have a distributed control system, where each subsystem uses its own state as well as the states of all other physically coupled subsystems. In other words, the control network is structurally the same as the plant network. As the third alternative, the structure matrix Γ can generate a middle-of-the-road solution, between distributed control approaches and decentralized ones, $\Gamma \subseteq \mathsf{S}(A)$, regarded as sparsely distributed control systems. This can be beneficial when some constraints on communication requirements between local controllers exist and hence the control network cannot have the same structure as the plant network. Besides, one may resort to optimizing the structure of the control network. As an illustration, a number of works in the literature focus on finding the sparsest control network that satisfies a global control objective; see e.g. [88]. This issue will be subject of the next section.

Remark 7.2 *The control network should always be a subset of the dynamics network, that is, $\Gamma \subseteq \mathsf{S}(A)$. In other words, if $A_{ij} = 0$ (subsystem j does not influence i-th subsystem), then $\gamma_{ij} = 0$.*

7.3 Optimal Structured SMC Design Problem

This section aims to design sliding matrices S_i while ensuring overall stability and penalizing the level of required control effort to maintain sliding as well as the stability of the reduced order interconnected systems. In order to cope with the above problem we may resort to selecting the switching function matrices S_i, with given λ_i, while ensuring overall stability and the stability of the reduced order interconnected systems, so that the linear control part of (7.17) minimizes the cost functional

$$J := \int_0^\infty \{x^T(\tau)Qx(\tau) + u^T(\tau)Ru(\tau)\}\, d\tau, \tag{7.18}$$

where $Q \in \mathbb{R}^{n\times n}$, with $n = \sum_{i=1}^h n_i$, is a given positive semi-definite matrix, and $R := \mathrm{diag}[R_i]_{i=1}^h \in \mathbb{R}^{m\times m}$, with $m = \sum_{i=1}^h m_i$, is a given block diagonal s.p.d matrix ($0 < R_i \in \mathbb{R}^{m_i\times m_i}$).

7.3.1 $\mathscr{H}_2$ based optimal structured static output feedback

Consider the controller in (7.17) contains only the linear part, hence

$$\begin{aligned} \dot{x}(t) &= Ax(t) + w(t) + Bu(t) \\ z(t) &= \tilde{C}_z x(t) + \tilde{D}_z u(t) \\ u(t) &= -\,\mathrm{diag}\left[\begin{bmatrix} M_i & I_{m_i}\end{bmatrix}\right]_{i=1}^h (\Gamma \circ A_\lambda)x(t), \end{aligned} \tag{7.19}$$

where $w(t) := \mathrm{col}(w_i(t))_{i=1}^h$ is a fictitious exogenous disturbance, $z(t) := \mathrm{col}(z_i(t))_{i=1}^h$, and

$$\tilde{C}_z := \begin{bmatrix} Q^{\frac{1}{2}} \\ 0 \end{bmatrix}, \quad \tilde{D}_z := \begin{bmatrix} 0 \\ R^{\frac{1}{2}} \end{bmatrix}. \tag{7.20}$$

In order to cope with the optimal SMC problem explained previously, this chapter then will endeavor to choose block diagonal matrix S so that the obtained closed-loop system by applying the linear control law in (7.19) minimizes

$$J := \|T_{wz}\|_2^2, \tag{7.21}$$

where $\|T_{wz}\|_2$ denotes the $\mathscr{H}_2$-norm of the closed loop transfer function from $w(t)$ to $z(t)$.

Remark 7.3 *It should be noted that designing the sliding surface with only the linear part of the controller is a standard scheme in the existing literature of SMC; see e.g. [116].*

The linear controller in (7.19) can be rewritten as $u(t) = Fx(t)$ in which

$$\begin{aligned} F &= -\,\mathrm{diag}\left[\begin{bmatrix} M_i & I_{m_i}\end{bmatrix}\right]_{i=1}^h (\Gamma \circ A_\lambda) \\ &= -\left(\mathrm{diag}\left[M_i\right]_{i=1}^h \mathrm{diag}\left[\begin{bmatrix} I_{n_i-m_i} & 0_{(n_i-m_i)\times m_i}\end{bmatrix}\right]_{i=1}^h - \mathrm{diag}\left[\begin{bmatrix} 0 & I_{m_i}\end{bmatrix}\right]_{i=1}^h\right)(\Gamma \circ A_\lambda). \end{aligned} \tag{7.22}$$

As seen M_i are the design freedoms in this new framework. Let us obtain the closed-loop system as

$$\begin{aligned}
A+BF =& A - \mathrm{diag}\left[\begin{bmatrix} 0 & 0 \\ 0 & I_{m_i}\end{bmatrix}\right]_{i=1}^{h} (\Gamma \circ A_\lambda) \\
& - B\mathrm{diag}\left[M_i\right]_{i=1}^{h} \mathrm{diag}\left[\begin{bmatrix} I_{n_i-m_i} & 0_{(n_i-m_i)\times m_i}\end{bmatrix}\right]_{i=1}^{h} (\Gamma \circ A_\lambda) \\
\triangleq & A_c + BMC,
\end{aligned} \tag{7.23}$$

where $M = \mathrm{diag}\left[M_i\right]_{i=1}^{h}$ and

$$\begin{aligned}
A_c &= A - \mathrm{diag}\left[\begin{bmatrix} 0 & 0 \\ 0 & I_{m_i}\end{bmatrix}\right]_{i=1}^{h} (\Gamma \circ A_\lambda), \\
C &= -\mathrm{diag}\left[\begin{bmatrix} I_{n_i-m_i} & 0_{(n_i-m_i)\times m_i}\end{bmatrix}\right]_{i=1}^{h} (\Gamma \circ A_\lambda).
\end{aligned} \tag{7.24}$$

Now consider the fictitious system

$$\begin{aligned}
\dot{x}(t) &= A_c x(t) + w(t) + B\bar{u}(t) \\
z(t) &= C_z x(t) + D_z \bar{u}(t) \\
y(t) &= Cx(t),
\end{aligned} \tag{7.25}$$

where $\bar{u}(t) = My(t)$ and

$$C_z = \begin{bmatrix} Q^{\frac{1}{2}} \\ -R^{\frac{1}{2}}\mathrm{diag}\left[\begin{bmatrix} 0 & I_{m_i}\end{bmatrix}\right]_{i=1}^{h} (\Gamma \circ A_\lambda)\end{bmatrix}, \; D_z = \begin{bmatrix} 0 \\ R^{\frac{1}{2}}\end{bmatrix}. \tag{7.26}$$

From this new viewpoint, the problem of designing $\mathscr{H}_2$ state feedback SMC (7.19)-(7.21) can be regarded as a *static output feedback LQ problem* for the fictitious system (A_c, B, C), given in (7.25). Specifically, minimizing the $\mathscr{H}_2$-norm of the T_{wz} (see (7.21)) subject to (7.19) is equivalent to minimizing the $\mathscr{H}_2$-norm of (7.25) with respect to the static output feedback gain M.
Different methods have been proposed in the literature to deal with the static output feedback LQ problem in (7.25), e.g. the iterative LMI method proposed in [71] referred to as the scaled min-max algorithm. We adopt the method in [71] (see Algorithm 7.1 in Section 7.6) to find M and thus S.

Remark 7.4 *Notice that the value of the $\mathscr{H}_2$ cost obtained from Algorithm 7.1 is not the true one, due to the conservatism introduced by assuming the block-diagonal structure for M. Nevertheless, the true value can be computed by solving the following Lyapunov equation*

$$P_{true}(A+BF) + (A+BF)^T P_{true} + Q + F^T RF = 0. \tag{7.27}$$

Then one can find the $\mathscr{H}_2$ cost as $\sqrt{trace(P_{true})}$.

7.3.2 Stability analysis of sliding mode dynamics

It should be noted that the $\mathscr{H}_2$ based method presented in the previous subsection may not necessarily stabilize the sliding mode dynamics. This subsection aims to impose an additional reduced order stability constraint on the previously proposed optimization problem. Let us rewrite the system in (7.1) as

$$\begin{bmatrix}\dot{x}_{i1}(t)\\ \dot{x}_{i2}(t)\end{bmatrix} = \begin{bmatrix}A_{i11} & A_{i12}\\ A_{i21} & A_{i22}\end{bmatrix}\begin{bmatrix}x_{i1}(t)\\ x_{i2}(t)\end{bmatrix} + \sum_{j=1,\ j\neq i}^{h}\begin{bmatrix}A_{ij11} & A_{ij12}\\ A_{ij21} & A_{ij22}\end{bmatrix}\begin{bmatrix}x_{j1}(t)\\ x_{j2}(t)\end{bmatrix} + B_i[u_i(t)+f_i(x_i)]. \tag{7.28}$$

Now by applying the equivalent control:

$$u_{eq,i} = -\begin{bmatrix}M_i & I_{m_i}\end{bmatrix}\left\{A_{\lambda,i}x_i(k) + \sum_{j=1}^{h}A_{ij}x_j(t)\right\} - f_i(x_i), \tag{7.29}$$

and using the nonsingular coordinate transformations $T = \text{diag}[T_i]_{i=1}^{h}$ with $T_i = \begin{bmatrix} I & 0\\ M_i & I\end{bmatrix}$, in the new coordinates, i.e. $\bar{x} = Tx$, we can write

$$\begin{bmatrix}\dot{x}_{i1}(t)\\ \dot{\sigma}_i(t)\end{bmatrix} = \begin{bmatrix}\bar{A}_{i11} & A_{i12}\\ 0 & \lambda_i I_{m_i}\end{bmatrix}\begin{bmatrix}x_{i1}(t)\\ \sigma_i(t)\end{bmatrix} + \sum_{j=1,\ j\neq i}^{h}\begin{bmatrix}\bar{A}_{ij11} & A_{ij12}\\ 0 & 0\end{bmatrix}\begin{bmatrix}x_{j1}(t)\\ \sigma_j(t)\end{bmatrix}, \tag{7.30}$$

where $\bar{A}_{i11} = A_{i11} - A_{i12}M_i$ and $\bar{A}_{ij11} = A_{ij11} - A_{ij12}M_j$. Obviously, (7.30) includes a reduced order interconnected system composed of h subsystems with dimension $n_i - m_i$. Note that this reduced order system is the same as the reduced order system resulted by the SMC (7.17) as they both have the same sliding surface. Therefore the stability of the system (7.30) will infer the stability of the reduced order system in Section 2, thus guaranteeing the stabilizing of the proposed SMC. Now, a stability analysis is considered for the system (7.30). Let the overall closed-loop system, obtained by the overall equivalent control, be

$$\dot{\bar{x}}(t) = \bar{A}^r_{cl}\bar{x}(t). \tag{7.31}$$

It can readily be shown that the stability of $\bar{A}^r_{cl}$ is equivalent to the stability of $A^r_{cl} = A^r_c + BMC^r$, where A^r_c and C^r are obtained from (7.24) by letting $\Gamma = \mathbb{1}_{h\times h}$, that is no structure imposed. Now in order to ensure the stability of the sliding mode dynamics, we augment the $\mathscr{H}_2$ problem in (7.41) (see Section 7.6) by including (7.42) with an s.p.d decision variable $\bar{P} > 0$. It is not hard to show that the obtained $M = \text{diag}[M_i]_{i=1}^{h}$ ensures the stability of the following composite reduced order dynamics:

$$\dot{x}_r(t) = \begin{bmatrix}\bar{A}_{111} & \cdots & \bar{A}_{1h11}\\ \vdots & \ddots & \vdots\\ \bar{A}_{h111} & \cdots & \bar{A}_{h11}\end{bmatrix}x_r(t) \tag{7.32}$$

where $x_r = \text{col}(x_{i1})_{i=1}^{h}$. Note that the obtained switching function matrices S_i are completely determined by choice of M_i.

We finally summarize the proposed structured $\mathscr{H}_2$ based SMC in the following theorem.

Theorem 7.2 *Assume that Algorithm 7.1, with a given structure matrix* Γ*, has a solution* $M = diag[M_i]_{i=1}^h$ *for some* $\delta > 0$*. Then the* $\mathscr{H}_2$ *performance constraint* $\|T_{wz}\|_2^2 < \delta$ *on the system* (7.19) *is ensured. After the reaching time* t_s*, the resulting reduced* $n_i - m_i$ $(i = 1, \cdots, h)$ *order sliding mode dynamics, obtained by applying the control law in* (7.11) *and* (7.10) *to the system* (7.1)*, is asymptotically stable.*

Proof 20 *The proof is trivial from the previously given method to select the sliding function matrix.*

7.4 Sparsification of the Control Network

Previous sections have studied the problem of designing $\mathscr{H}_2$-based SMC for NCSs with imposing *a priori* constraints on communication requirements among subsystems. In other words, the structure matrix Γ in (7.17) is assumed to be known a priori. The objective in this section is to establish an optimization framework which indeed aims to obtain a trade-off between the $\mathscr{H}_2$ performance and the sparsity of the control network structure. Indeed, one can say that the main objective here is to minimize the cost of the control network utilized to control (stabilize) the system. Here, we assume that the general costs, including the construction and data transferring costs etc, are identical for all the links. Hence, the minimization of control network costs can intuitively be considered as the minimization of the number of links in the control network structure or equivalently finding the sparsest control network structure that can satisfy a global control objective. On the other hand, minimizing the control network structure for the SMC without taking into account the control costs, may not result in applicable results. Here we propose a way to minimize the control performance and the communication costs simultaneously. We formulate this problem as

$$\begin{aligned} &\min \;\; J(\Gamma, M) + \eta\, \mathbf{card}(\Gamma) \\ &\text{subject to } \Gamma \subseteq \mathsf{S}(A),\ \mathsf{S}(M) = I, \text{ and } (7.25), \end{aligned} \tag{7.33}$$

where J is the square of the $\mathscr{H}_2$ norm of the closed-loop transfer function from $w(t)$ to $z(t)$ in (7.25), $\Gamma = [\gamma_{ij}]_{h \times h}$ and $\mathbf{card}(\cdot)$ denotes the cardinality function (the number of nonzero elements of a matrix). Besides, $\eta \geq 0$ is a given constant which captures a trade-off between the $\mathscr{H}_2$ performance and the sparsity of the controller structure. For example a larger η will lead to a sparser Γ and $\eta = 0$, which means $\Gamma = \mathsf{S}(A)$, converts the problem to a distributed SMC with the objective function (7.21). The optimization problem in (7.33) is a mixed-binary problem which, broadly speaking, requires an intractable combinatorial search to achieve the solution.

Notice that the cardinality function, in optimization problems such as (7.33), is usually approximated by the ℓ_1 norm of the optimization variable [30] or the so-called weighted ℓ_1 norm [31]. Since the weighted ℓ_1 norm is not implementable (the required weights should be calculated based on the unknown feedback gain), a reweighted algorithm is proposed in [31], and further used by [88] to design sparse feedback gains. This algorithm solves weighted optimization problems iteratively in which the weights are updated inversely proportional to the strength of individual (block) entries of feedback gain in the previous iteration. However, the existing reweighted algorithms are not applicable to the optimization problem in (7.33), as the system matrices A_c and C in the fictitious system (7.25), involve the structure matrix Γ. Instead, in this chapter, we will consider a heuristic scheme by relaxing the constraint on the variables γ_{ij}, $i \neq j$ from the binary variables, 0 or 1, to the constraint of $0 \leq \gamma_{ij}^r \leq 1$, $i \neq j$, where

$$\gamma_{ij}^r = \frac{\|F_{ij}\|_F}{\max\limits_{\gamma_{ij}=1} \|F_{ij}\|_F} \qquad i \neq j, \tag{7.34}$$

in which F_{ij} denotes the ij-th entry of the control feedback gain F in (7.22) and $\|\cdot\|_F$ is the Frobenius norm. Indeed γ_{ij}^r can be considered as the normalized strength of the coupling feedback gain F_{ij}. This scheme works by first finding the normalized strengths of all the coupling feedbacks and then removing the links corresponding to the weaker feedback gains one-by-one until the stability of the overall closed-loop system is violated. Indeed, by assigning a normalized weight to each link according to the contribution of its corresponding feedback gain in the control objective, this process will reduce the probability of losing the stability by removing a link. This also can lead to a more computationally efficient method compared to an exhaustive search without taking into account the strength of the coupled feedback gains.

Procedure 7.1

1) *Initialize* $\Gamma = \mathsf{S}(A)$ *and* $l = 1$, *in which* l *denotes the iteration number.*

2) *Solve Algorithm 7.1 (refer to Section 7.6) to find* P *and* M. *If the LMIs in* (7.41) *and* (7.42) *are feasible,* $\Gamma^l \leftarrow \Gamma$ *and* $J_s(l) = J(\Gamma^l) + \eta\mathbf{Card}(\Gamma^l)$, *otherwise terminate the search and the problem has no solution.*

3) *Find* γ_{ij}^r *as in* (7.34) *for all* $\gamma_{ij} = 1$, $i \neq j$. *Sort the set* $\{\gamma_{ij}^r\}$ *in ascending order.*

4) *Set* γ_{ij} *corresponding to the* l*-th entry of* $\{\gamma_{ij}^r\}$ *to zero and* $l = l + 1$.

5) *Solve Algorithm 7.1. If the LMIs in* (7.41) *and* (7.42) *are feasible,* $\Gamma^l \leftarrow \Gamma$, *then compute the objective function in* (7.33) *to find* $J_s(l) = J(\Gamma^l) + \eta\mathbf{Card}(\Gamma^l)$, *if* $l \leq \mathbf{Card}(\mathsf{S}(A))$, *return to Step 4, otherwise go to Step 6.*

6) *Find* $l^\star = \arg\min\limits_{l} J_s(l)$ *and return its corresponding* $\Gamma^{l^\star}$.

Remark 7.5 *It should be noted that random truncation of the distributed controller* ($\Gamma = \mathsf{S}(A)$) *may lead to a feedback that cannot stabilize the overall system. In contrast, the proposed method here is a systematic way to reduce the number of links in the control network structure while preserving the stability of the overall closed-loop system.*

Notice that in order to obtain the result from the above procedure, at most $\sum_{l=1}^{\mathscr{A}} \mathscr{E}_l$ convex problems need to be solved, where $\mathscr{A} = \mathbf{Card}(\mathsf{S}(A))$ and $\mathscr{E}_l$ denotes the number of iterations that is required for Algorithm 7.1 at the l-th iteration of Procedure 7.1. This is in contrast to $\sum_{i=1}^{2^{\mathscr{A}}} \mathscr{E}_i$ in the case of carrying an exhaustive search on the binary variables. Besides, roughly speaking, all the methods in the literature to solve an $\mathscr{H}_2$ static output feedback problem utilize iterative processes, and their solutions and more importantly their convergence depend quite significantly on the initial conditions. It is difficult to ensure Procedure 7.1 to achieve the global minimum or even a local one. However, our extensive computational experiments show that this algorithm can provide an effective means to achieve an acceptable trade-off between the control performance and the sparsity of the control network structure. Compared to the exhaustive search, Procedure 7.1 proposes a simple suboptimal relaxation scheme, which is much more computationally attractive.

7.5 Numerical Examples

7.5.1 Example 1

Once again we consider the interconnected system in Section 6.6.1 including three inverted pendulums mounted on coupled carts [114, 104]. The linearized system equations are also given in Section 6.6.1. Define $x_i = [x_{i,1}, x_{i,2}, x_{i,3}, x_{i,4}]^T = [\theta_i, \dot{\theta}_i, \mathbf{x}_i, \dot{\mathbf{x}}_i]^T$,

$$A_i = \begin{bmatrix} 0 & 1 & 0 & 0 \\ \frac{M_i+m}{M_i\ell} g & 0 & \frac{k_i}{M_i\ell} & \frac{c_i+b_i}{M_i\ell} \\ 0 & 0 & 0 & 1 \\ \frac{-m}{M_i} g & 0 & \frac{-k_i}{M_i} & \frac{-c_i-b_i}{M_i\ell} \end{bmatrix}, \quad A_{ij} = \begin{bmatrix} 0 & 0 & 0 & 0 \\ 0 & 0 & \frac{-k_{ij}}{M_i\ell} & \frac{-b_{ij}}{M_i\ell} \\ 0 & 0 & 0 & 0 \\ 0 & 0 & \frac{k_{ij}}{M_i} & \frac{b_{ij}}{M_i} \end{bmatrix}$$

$$B_i = \begin{bmatrix} 0 & \frac{-1}{M_i\ell} & 0 & \frac{1}{M_i} \end{bmatrix}^T,$$

for $(i, j) \in \{(1,2), (2,1), (2,3), (3,2)\}$, in which $k_i = \sum_{j \in \mathsf{J}_i} k_{ij}$ and $b_i = \sum_{j \in \mathsf{J}_i} b_{ij}$, where $\mathsf{J}_i := \{j \mid [\mathsf{S}(A)]_{ij} = 1, j \neq i\}$. Besides, c_i, $b_{ij} = b_{ji}$, $k_{ij} = k_{ji}$ and ℓ are friction, damper, spring coefficients and pendulum length respectively. It is also assumed that the moment of inertia of each pendulum is zero. Besides, the system parameters are assumed as $M_1 = 4$, $M_2 = 3$, $M_3 = 5$, $m = 0.2$, $g = 10$, $\ell = 4$, $k_{12} = k_{21} = 1$, $k_{23} = k_{32} = 1$, $b_{12} = b_{21} = 1$, $b_{23} = b_{32} = 0.2$, $c_1 = 0.4$, $c_2 = 0.2$ and $c_3 = 0.1$. Notice that here it is assumed that entire system states are available. Transformation matrices T_i are utilized to transform the subsystems to the form given in (7.2). The performance weights are set as $Q = I_n$ and $R = 0.1 I_m$. We have also chosen $\lambda = -1$. Procedure 7.1, with three different parameters $\eta = 0.001$, $\eta = 0.002$ and $\eta = 0.01$, is solved and the corresponding results are as follows. When $\eta = 0.001$ Procedure 7.1 suggests removing all the links between subsystems except links among subsystem 1 and subsystem 2 (i.e. $\gamma_{12} = \gamma_{21} = 1$ and $\gamma_{13} = \gamma_{31} = \gamma_{23} = \gamma_{32} = 0$). Also with $\eta = 0.002$, this procedure promotes the same structure. Eventually, Procedure 7.1, with $\eta = 0.01$ suggests removing all the links between subsystems (i.e. $\gamma_{ij} = 0,\ \forall\ i, j = 1,2,3$). Furthermore, the initial conditions for Algorithm 7.1, which is required to be solved for addressing the suboptimal LQ static output feedback problem in Section 7.3, are set to $\mu_{sol} = 1$, $Y_{sol} = I_n$ and $\bar{Y}_{sol} = I_n$ and the parameter $\delta = 180$. As seen, the larger parameter η results in a more sparse control network.

7.5.2 Example 2

Consider the system (7.1) with the following parameters:

$$A = \left[\begin{array}{ccc|cc|cc} 0 & 0 & -3.6 & 0 & 0 & 0 & 0 \\ -0.2 & 7.2 & -0.4 & -0.1 & 0 & 0.3 & 0 \\ 0.3 & 0 & 0 & 3.0 & 0.2 & 0 & 0 \\ \hline 0 & 0.3 & 0 & 0 & 1.0 & 0 & 0 \\ 0 & 0 & 0 & 0 & 0 & 0 & -0.2 \\ \hline 0 & 0 & 0 & 0 & 0 & 0 & 1.0 \\ 0 & 0.3 & 0 & -0.5 & 0.1 & 0 & 0 \end{array}\right], \quad B = \left[\begin{array}{cc|c|c} 0 & 0 & 0 & 0 \\ 1 & 0 & 0 & 0 \\ 0 & 1 & 0 & 0 \\ \hline 0 & 0 & 0 & 0 \\ 0 & 0 & 1 & 0 \\ \hline 0 & 0 & 0 & 0 \\ 0 & 0 & 0 & 1 \end{array}\right].$$

Note that all three open-loop local subsystems are unstable. The performance weights Q and R are set to identity matrices and we choose $\lambda = -4$. We first use Procedure 7.1 with three different parameters $\eta = 0.001$, $\eta = 0.01$ and $\eta = 0.1$, and the achieved results are as follows When $\eta = 0.001$ Procedure 7.1 suggests keeping all the links between subsystems (i.e. $\gamma_{ij} = 0,\ \forall\ i, j = 1,2,3$). Moreover, with $\eta = 0.01$, the procedure proposes keeping only the link between subsystems 1 and 2 (i.e. $\gamma_{12} = 1$ and $\gamma_{21} = \gamma_{13} = \gamma_{31} = \gamma_{23} = \gamma_{32} = 0$). Eventually, Procedure 7.1, with $\eta = 0.1$ suggests removing all the links between subsystems (i.e. $\gamma_{ij} = 0,\ \forall\ i, j = 1,2,3$). The initial conditions for Algorithm 7.1, which solves the suboptimal LQ static output feedback problem, are $\mu_{sol} = 0.1$, $Y_{sol} = I_n$ and $\bar{Y}_{sol} = I_n$ and the parameter $\delta = 190$. One can see that as the regularization parameter η increases, the control network becomes more sparse.

7.5.2.1 Comparison 1

For comparison let us consider a standard sparse state feedback LQR controller with the given choices of Q and R. In doing so, assume that there exists a stabilizing $\Gamma \circ F_{lqr}$ with $\Gamma = [\gamma_{ij}]_{h\times h}$ and $F_{lqr} \in \mathbb{R}^{m\times n}$ minimizing the following cost functional,

$$J = \text{trace}(\bar{X}_{lqr}), \tag{7.35}$$

where $\bar{X}_{lqr} = \text{diag}[\bar{X}_i]_{i=1}^{h} > 0$ is obtained from solving the Lyapunov inequality,

$$\begin{aligned}[A + B(\Gamma \circ F_{lqr})]^T \bar{X}_{lqr} + \bar{X}_{lqr}[A + B(\Gamma \circ F_{lqr})] \\ + Q + (\Gamma \circ F_{lqr})^T R (\Gamma \circ F_{lqr}) < 0. \end{aligned} \tag{7.36}$$

The above inequality is not convex with respect to $\bar{X}_{lqr}$ and $\Gamma \circ F_{lqr}$. However, it can be convexified through variable changing. Letting $X_{lqr} = \bar{X}_{lqr}^{-1}$ and pre and post multiplying X_{lqr} to (7.36), we have

$$X_{lqr}[A+B(\Gamma\circ F_{lqr})]^T+[A+B(\Gamma\circ F_{lqr})]X_{lqr}+X_{lqr}QX_{lqr} \\ +X_{lqr}(\Gamma\circ F_{lqr})^T R(\Gamma\circ F_{lqr})X_{lqr}<0. \tag{7.37}$$

Having the convex constraint in (7.37), the minimization problem explained in (7.35) can be cast as an optimization problem utilizing the LMI approach,

$$\text{minimize} \quad \text{trace}\ (Z_s) \quad \text{subject to}$$

$$\begin{bmatrix} AX_{lqr}+X_{lqr}A^T+BY_{lqr}+Y_{lqr}^T B^T & \star & \star \\ Q^{\frac{1}{2}}X_{lqr} & -I & \star \\ R^{\frac{1}{2}}Y_{lqr} & 0 & -I \end{bmatrix}<0, \tag{7.38}$$

$$\begin{bmatrix} -Z_s & \star \\ I & -X_{lqr} \end{bmatrix}<0, \tag{7.39}$$

where $Y_{lqr}=(\Gamma\circ F_{lqr})X_{lqr}=\Gamma\circ(F_{lqr}X_{lqr})=\Gamma\circ\bar{Y}_{lqr}$ with $\bar{Y}_{lqr}\in\mathbb{R}^{m\times n}$, and Z_s is a slack variable. Thus, the structural state feedback can be obtained as $\Gamma\circ F_{lqr}=Y_{lqr}X_{lqr}^{-1}$.

Remark 7.6 *It is easy to realize that*

$$\mathsf{S}(X_{lqr}^{-1})=\mathsf{S}(X_{lqr})=I,$$

and since $\mathsf{S}(Y_{lqr})\subseteq\Gamma$, *thus*

$$\mathsf{S}(Y_{lqr}X_{lqr}^{-1})\subseteq\Gamma.$$

This means that the structural state feedback gain $\Gamma\circ F_{lqr}$ *obtained from* $Y_{lqr}X_{lqr}^{-1}$ *has the desired structure* Γ.

Solving the minimization problem in (7.38) and (7.39) gives a bound of 6.3837 on the $\mathscr{H}_2$ cost for decentralized structure. However, again it should be noted that due to the conservatism introduced by enforcing a block-diagonal structure on X_{lqr}, this is not the true value of $\mathscr{H}_2$ cost and it can be obtained as 5.8276 from solving the Lyapunov equation in (7.27) with the resulting F_{lqr}. Notice also that the true value of $\mathscr{H}_2$ cost achieved from Algorithm 7.1 for decentralized structure is 6.3080.

7.5.2.2 Comparison 2

We now consider a reweighted ℓ_1 algorithm for finding an *optimal sparse state feedback* gains; e.g. see [88]. This problem can be cast as follows:

$$\text{Minimize trace}(Z_s) + \bar{\eta} \left\| W \circ V_{lqr} \right\|_{\ell_1} \tag{7.40}$$

subject to

$$\begin{bmatrix} AX_{lqr} + X_{lqr}A^T + BV_{lqr} + V_{lqr}^T B^T & \star & \star \\ Q^{\frac{1}{2}} X_{lqr} & -I & \star \\ R^{\frac{1}{2}} V_{lqr} & 0 & -I \end{bmatrix} < 0,$$

$$\begin{bmatrix} -Z_s & \star \\ I & -X_{lqr} \end{bmatrix} < 0,$$

where $0 < X_{lqr} \in \mathbb{R}^{n \times n}$, $V_{lqr} \in \mathbb{R}^{m \times n}$, Z_s is a slack variable and W is a given weighting matrix with the same dimension of $\mathsf{S}(F_{lqr})$. We then exploit Algorithm 7.2 in Section 7.7 with $\bar{\eta} = 0.01$ to find the sparse state feedback matrix. This algorithm suggests the decentralized structure for the control network with a true value $\mathscr{H}_2$ cost of 5.8276. Also, notice that the existing reweighted ℓ_1 algorithm, for minimizing the network structure, is not applicable to our problem which was indeed the design of a sparse distributed $\mathscr{H}_2$-based SMC and rearranged as an *LQ* static output feedback problem.

7.6 Algorithm for Solving the *LQ* Static Output Feedback Problem

Consider the system in (7.25). As mentioned the objective is to design $M = \text{diag}[M_i]_{i=1}^h$ so that the $\mathscr{H}_2$ norm from $w(t)$ to $z(t)$ is less than a given constant δ while the stability of the composite sliding mode dynamics is ensured. According to e.g. [71], this problem can be cast as finding two symmetric $P > 0$ and $\bar{P}$ such that

$$\begin{aligned} &PA_{cl} + A_{cl}^T P + C_{cl}^T C_{cl} < 0 \\ &\text{trace}(P) < \delta, \end{aligned} \tag{7.41}$$

$$\bar{P}A_{cl}^r + (A_{cl}^r)^T \bar{P} < 0, \tag{7.42}$$

in which $A_{cl} = A_c + BMC$, $A_{cl}^r = A_c^r + BMC^r$ and $C_{cl} = C_z + D_z MC$, and A_c^r and C^r are defined in Section 7.3.2. To deal with this problem, [71] proposes the so-called iterative scaled min-max method. To explain this method, we need to introduce four scalar variables ν, β, ψ, ϖ and four symmetric matrices $0 < X \in \mathbb{R}^{n \times n}$, $0 < Y \in \mathbb{R}^{n \times n}$, $0 < \bar{X} \in \mathbb{R}^{n \times n}$ and $0 < \bar{Y} \in \mathbb{R}^{n \times n}$. Now the scaled min-max algorithm can be summarized as follows.

Algorithm 7.1 1) *Initialize* Y_{sol}, $\bar{Y}_{sol}$, $\beta_{sol} > 0$ *and set* $\varepsilon > 0$ *(termination scalar),* $l = 1$ *(iteration number).*

2) *Solve*

$$\begin{aligned}
&\min_{X,\bar{X},\varpi} \quad \psi_l \\
&I \le Y_{sol}^{\frac{1}{2}} X Y_{sol}^{\frac{1}{2}} \le \psi_l I \\
&I \le \bar{Y}_{sol}^{\frac{1}{2}} \bar{X} \bar{Y}_{sol}^{\frac{1}{2}} \le \psi_l I \\
&1 \le \varpi \beta_{sol} \le \psi_l \\
&\begin{bmatrix} B \\ D_z \end{bmatrix}^{\perp} \begin{bmatrix} A_c X + X A_c^T & X C_z^T \\ C_z X & -\varpi I \end{bmatrix} \begin{bmatrix} B \\ D_z \end{bmatrix}^{\perp T} \le -I \\
&B^{\perp} \left(A_c^r \bar{X} + \bar{X} \left(A_c^r\right)^T \right) B^{\perp T} \le -I,
\end{aligned}$$

to find X_{sol}, $\bar{X}_{sol}$ *and* ϖ_{sol}.

3) *With given* X_{sol}, $\bar{X}_{sol}$ *and* ϖ_{sol} *solve*

$$\begin{aligned}
&\max_{Y,\bar{Y},\beta} \ \nu_l \\
&\nu_l I \le X_{sol}^{\frac{1}{2}} Y X_{sol}^{\frac{1}{2}} \le I \\
&\nu_l I \le \bar{X}_{sol}^{\frac{1}{2}} \bar{Y} \bar{X}_{sol}^{\frac{1}{2}} \le I \\
&\nu_l \le \varpi_{sol} \beta \le 1 \\
&\begin{bmatrix} C^{T\perp}(Y A_c + A_c^T Y + \beta C_z^T C_z) C^{T\perp T} & \beta I \\ \beta I & -I \end{bmatrix} \le 0 \\
&\begin{bmatrix} trace(Y) - \delta\beta & \beta \\ \beta & -I \end{bmatrix} \le 0 \\
&C^{T\perp} \left(\bar{Y} A_c^r + \left(A_c^r\right)^T \bar{Y} \right) C^{T\perp T} \le 0,
\end{aligned}$$

to find Y_{sol} *and* β_{sol}.

4) *If* $\lambda_{\min}(Y_{sol}) < \varepsilon$ *or* $\lambda_{\min}(\bar{Y}_{sol}) < \varepsilon$ *or* $\beta_{sol} < \varepsilon$ *then stop, the algorithm does not converge.*

5) *If* $\psi_l - \nu_l < \varepsilon$*, go to Step 6, otherwise* $l = l + 1$ *and return to Step 2.*

6) *Return* $P = \beta^{-1} Y$ *and* $\bar{P} = \beta^{-1} \bar{Y}$.

If the algorithm converges to the solution, then $\psi_l \to 1$, $\nu_l \to 1$, $X \to Y^{-1}$, $\bar{X} \to \bar{Y}^{-1}$ and $\beta \to \varpi^{-1}$. The required M then can be obtained by solving (7.41) and (7.42) with given P and $\bar{P}$.

7.7 Reweighted ℓ_1 Minimization Algorithm

Using the reweighted ℓ_1 norm for promoting sparsity has been considered in e.g. [44, 30]. Define the matrix $N = \mathbb{1}_{m\times n}$. It can be shown that (e.g. see [44, 30]) the optimization problem in (7.40) is equivalent to

$$\begin{aligned}
&\text{Minimize } \operatorname{trace}(Z_s) + \bar{\eta}\operatorname{trace}(N^T G) \\
&\text{subject to} \\
&\begin{bmatrix} AX_{lqr} + X_{lqr}A^T + BV_{lqr} + V_{lqr}^T B^T & \star & \star \\ Q^{\frac{1}{2}} X_{lqr} & -I & \star \\ R^{\frac{1}{2}} V_{lqr} & 0 & -I \end{bmatrix} < 0, \\
&\begin{bmatrix} -Z_s & \star \\ I & -X_{lqr} \end{bmatrix} < 0, \\
&-G \leq W \circ V_{lqr} \leq G,
\end{aligned} \tag{7.43}$$

where $0 < X_{lqr} \in \mathbb{R}^{n\times n}$, $V_{lqr} \in \mathbb{R}^{m\times n}$, Z_s is a slack variable, W denotes the weighting matrix and the last inequality is element-wise with $G \in \mathbb{R}^{m\times n}$ whose entries are nonnegative. Then the algorithm to solve the above optimization problem is as the following,

Algorithm 7.21) *With given* $\varepsilon > 0$, $\alpha > 0$ *and* $\bar{\eta} > 0$, *initialize* $W = \mathbb{1}_{h\times h}$, $l = 1$ *and* $V^l = 0$.

2) *Solve the minimization problem* (7.43) *to obtain* $F_{lqr}^\star = V_{lqr}^\star X_{lqr}^{\star^{-1}}$.

3) *Update* $W_{ij} = \frac{1}{\left\|\left(V_{lqr}^\star\right)_{ij}\right\|_F + \varepsilon}$ *and form* $W = [W_{ij}]_{h\times h}$.

5) *If* $\left\|V_{lqr}^\star - V^l\right\| \leq \alpha$ *go to Step 6, else* $V^l = V_{lqr}^\star$, $l = l+1$ *and return to Step 2.*

6) *Return* $F_{lqr}^\star$.

Solving Algorithm 7.2 gives the most effective sparse structure of F_{lqr}. Then, by ignoring the unnecessary entries in F_{lqr} we find the structure matrix Γ. Eventually, the optimal structured feedback matrix is obtained by solving the problem in (7.38) and (7.39). This procedure is considered in e.g. [44].

7.8 Conclusions

This chapter has developed a distributed sliding mode control framework by using (some) other subsystems' states. Indeed this issue has been considered to widen the applicability region of the decentralized SMC in which each subsystem's controller uses only local information. Furthermore, an approach is proposed for the sliding surface design in which the level of required control effort is taken into account. Then this novel scheme has been utilized to present a heuristic algorithm that provides an effective means of selecting an overall sliding manifold through a trade-off between the performance and the sparsity of the controller. Indeed, the novel scheme proposed here to design the sliding surface helps avoid excessively large control effort. Illustrative examples have been used to demonstrate the effectiveness of the proposed approach.

Part IV

DSMC for Two-Dimensional Systems

8

Discrete-time SMC for two-dimensional systems

CONTENTS

8.1	Introduction	161
8.2	Problem Formulation	162
	8.2.1 New 1D form of 2D first FM model	163
8.3	DSMC for 1D Discrete Vector Form	165
	8.3.1 Direct method to find control law	168
8.4	Simulation Results	168
8.5	Conclusions	168

Abstract– In this chapter, a new approach (ID vectorial form) is introduced to set the stage for extending DSMC to two-dimensional (2D) systems. Using one-dimensional (1D) form to represent 2D systems can be used as an alternative strategy to reduce the inherent complexity of 2D systems and their applications. Unlike the Wave Advanced Model (WAM) form (proposed by Porter and Aravena), the suggested 1D vectorial form, in this chapter, has invariable dimension and hence a regular form of system can be derived which is the basis of discrete-time sliding mode control here. In this chapter, the first Fornasini and Marchesini (FM) model of 2D systems which is a second order recursive form is considered.

8.1 Introduction

Multidimensional linear systems and in particular 2D systems have attracted much attention since the 1970s; see [54, 53, 47, 46] for two-dimensional linear models. In 1972, *Givone* and *Roesser*, for the first time, introduced a state-space model for a linear iterative circuit which is studied as a spatial system rather than a temporal system [54], [53]. This state-space model is then referred to as the GR model. *Fornasini* and *Marchesini* proposed a different state-space realization for the 2D digital filters [47], known as the first FM model. Later in [46], they proposed a new state-space form which is the first-order difference equation and sometimes is called the second FM model. Since then, multidimensional systems, especially 2D systems, have been

studied in many aspects and in many applications.
In [143], according to the so-called 1D quasi-sliding mode ([68]), SMC design has been extended for 2D systems in Roesser Model (RM). In addition, the conditions to ensure the remaining horizontal and vertical states in RM on the switching surfaces and also the reaching condition for designing the control law using a 2D Lyapunov function are investigated in [3].

Another strategy to work with 2D systems is to transfer them to a 1D form. The wave advance model (WAM) is a 1D form of 2D systems established in [111]. From the view point of the WAM model, 2D systems are considered as advanced waves and consequently the original stationary 2D system is converted to a time-varying 1D system. Moreover, the system matrices are in rectangular form rather than square form. As a result, the major drawback of this 1D form of 2D systems is the varying dimensions of the defined state vectors. This means that the results developed using this framework are most likely computationally unattractive in terms of possible applications. Motivated by this issue and using stacking vectors, a new approach to converting 2D systems to a 1D form is proposed in this chapter. Specifically, here, rather than using the WAM model, a row (column) processing method is used. Row (column) processing means that the 2D variables which are in the same rows (columns) are used to form 1D stacking vectors. Consequently, the states, inputs and outputs of the obtained 1D system are in the vector form, and more importantly their dimensions are invariant. This framework is basically useful for a class of 2D linear systems in which information propagation in one of the two distinct directions only occurs over a finite horizon. This can be the case of a repetitive process [50] or any inherently 2D system, for instance, Darboux equation [73]:

$$\frac{\partial s(x,t)}{\partial x \partial t} = a_1 \frac{\partial s(x,t)}{\partial t} + a_2 \frac{\partial s(x,t)}{\partial x} + a_0 s(x,t) + bu(t,x),$$

where initial and boundary conditions are known and constant. Here, $s(x,t)$ is an unknown function at space ($x \in [0, x^\star]$) and $t \in \mathbb{Z}^+$, and all the coefficients are real scalars. $u(x,t)$ is the input signal. Defining $s(i\Delta x, j\Delta t) := s(i,j)$, and

$$\frac{\partial s(x,t)}{\partial x} \cong \frac{s(i+1,j) - s(i,j)}{\Delta x}, \quad \frac{\partial s(x,t)}{\partial t} \cong \frac{s(i,j+1) - s(i,j)}{\Delta t},$$

the discrete form of Darboux equation can be written as

$$\begin{aligned} s(i+1,j+1) =& (a_1\Delta t)s(i+1,j) + (a_2\Delta x)s(i,j+1) \\ &+ (a_0\Delta x\Delta t - a_1\Delta t - a_2\Delta x + 1)s(i,j) \\ &+ (b\Delta x\Delta t)u(i,j), \end{aligned}$$

where Δt and Δx denote the time step size and the spatial mesh size, respectively. As seen, discrete form of Darboux equation is a first FM model which has a finite propagation over the space direction.

8.2 Problem Formulation

To illustrate how 2D systems can be converted to 1D form, we consider the first FM model with the following formulation

$$x(i+1,j+1)=A_1x(i+1,j)+A_2x(i,j+1)+A_0x(i,j)+Bu(i,j) \tag{8.1}$$

where $x\in\mathbb{R}^n$ and $u\in\mathbb{R}^m$ are respectively local state and control input. In addition, the matrices in this equation are $A_1\in\mathbb{R}^{n\times n}$, $A_2\in\mathbb{R}^{n\times n}$, $A_0\in\mathbb{R}^{n\times n}$ and $B\in\mathbb{R}^{n\times m}$. It can be seen that this relation is a second order recursive equation.

8.2.1 New 1D form of 2D first FM model

The FM model (8.1) can be represented in the following form

$$x(i+1,j+1)-A_1x(i+1,j)=A_2x(i,j+1)+A_0x(i,j)+Bu(i,j). \tag{8.2}$$

Now, we define the following stacking vectors

$$V(i)=\begin{bmatrix} A_1x(i+1,0)+A_0x(i,0) \\ 0 \\ \vdots \\ 0 \end{bmatrix},$$

$$X(i)=\begin{bmatrix} x(i,1) \\ x(i,2) \\ \vdots \\ x(i,v) \end{bmatrix}, U(i)=\begin{bmatrix} u(i,0) \\ u(i,1) \\ \vdots \\ u(i,v-1) \end{bmatrix}, \tag{8.3}$$

where v is the dimension of distinct variable j, $X(i)\in\mathbb{R}^{v.n}$, $V(i)\in\mathbb{R}^{v.n}$ and $U(i)\in\mathbb{R}^{v.m}$. As a result, the 2D equation (8.2) can be presented as

$$JX(i+1)=KX(i)+LU(i)+V(i), \tag{8.4}$$

where

$$J=\begin{bmatrix} I & 0 & 0 & \dots & 0 & 0 \\ -A_1 & I & 0 & \dots & 0 & 0 \\ 0 & -A_1 & I & \dots & 0 & 0 \\ \vdots & \vdots & & \ddots & \vdots & \vdots \\ 0 & 0 & 0 & \dots & I & 0 \\ 0 & 0 & 0 & \dots & -A_1 & I \end{bmatrix}$$

$$=I_v\otimes I_n+\begin{bmatrix} 0_{1\times(v-1)} & 0 \\ I_{v-1} & 0_{(v-1)\times 1} \end{bmatrix}\otimes(-A_1),$$

$$K = \begin{bmatrix} A_2 & 0 & 0 & \dots & 0 & 0 \\ A_0 & A_2 & 0 & \dots & 0 & 0 \\ 0 & A_0 & A_2 & \dots & 0 & 0 \\ \vdots & \vdots & & \ddots & \vdots & \vdots \\ 0 & 0 & 0 & \dots & A_2 & 0 \\ 0 & 0 & 0 & \dots & A_0 & A_2 \end{bmatrix}$$

$$= I_v \otimes A_2 + \begin{bmatrix} 0_{1\times(v-1)} & 0 \\ I_{v-1} & 0_{(v-1)\times 1} \end{bmatrix} \otimes A_0. \tag{8.5}$$

Besides,

$$L = \begin{bmatrix} B & 0 & \dots & 0 \\ 0 & B & \dots & 0 \\ \vdots & \vdots & \ddots & \vdots \\ 0 & 0 & \dots & B \end{bmatrix} = I_v \otimes B.$$

Here, $x(i+1,0)$ and $x(i,0)$ are state boundary conditions on boundary ($j=0$). Moreover, as it is seen with the vectorial definition (8.3), the variable j is hidden in the new defined 1D form. Model (8.4) is also known as descriptor model.

Remark 8.1 *In general, the dimension of 2D systems can be infinite. However, as it was mentioned before, in this chapter, it is assumed that one of the distinct variables of 2D system is finite. Moreover, the computing limitations have made it inevitable to assume finite dimensions for both separate directions of 2D systems. In this report, the dimension of the considered 2D system is assumed to be $\mu \times v$ and, as a result, the size of 1D state vector $X(i)$ and control input vector $U(i)$ in* (8.4) *are v.n and v.m, respectively. Besides, there are two set of boundary conditions ($i=0$ and $j=0$).*

$$\begin{cases} \alpha(i) = x(i,0) & \textit{over } j=0, \\ \beta(j) = x(0,j) & \textit{over } i=0. \end{cases} \tag{8.6}$$

Remark 8.2 *Likewise, note that matrices J and K are bi-diagonal (in general block) Toeplitz matrices and the sizes of these matrices depend on the dimension of state vector $X(i)$. The dimensions of the matrices defined in* (8.4) *are*

$$\begin{aligned} &J : [v.n] \times [v.n], \ K : [v.n] \times [v.n], \\ &L : [v.n] \times [v.m]. \end{aligned} \tag{8.7}$$

To apply discrete sliding mode control to the system (8.4), this equation should be left multiplied by J^{-1} (obviously, matrix J is of full rank). In the case that the elements of matrix J are varying, in every step the inverse of this matrix should be computed. A very heavy computational load could result, especially for 2D grids with large dimensions. However, in our case, the matrix J is time invariant and consequently in the proposed DSMC of this chapter, the matrix J^{-1} can be computed only once.

Remark 8.3 *In [33], a simple formula for the inverse of a block matrix with non-zero* *blocks in the principal diagonal and the first sub-diagonal only is proved. Adapting this formula to our case results in the following form for* $J^{-1} = [\gamma_{p,q}]$,

$$\gamma_{p,q} = \begin{cases} 0_n & \text{if } p < q, \\ I_n & \text{if } p = q, \\ (-1)^{p+q}(-A_1)^{p-q} & \text{if } p > q. \end{cases} \tag{8.8}$$

Then, by left multiplying (8.4) by J^{-1}, the following standard 1D state space form can be obtained,

$$X(i+1) = \hat{K}X(i) + \hat{L}U(i) + \hat{R}V(i), \tag{8.9}$$

where

$$\hat{K} = J^{-1}K,\ \hat{L} = J^{-1}L, \text{ and } \hat{R} = J^{-1}.$$

Note that a numerical algorithm is given in [5] to compute J^{-1} explicitly. It can be found that $\hat{K}$ and $\hat{L}$ are block lower triangular matrices as

$$\hat{K} = \begin{bmatrix} A_2 & 0 & \cdots & 0 & 0 \\ A_1A_2 + A_0 & A_2 & \cdots & 0 & 0 \\ \vdots & \vdots & & \vdots & \vdots \\ A_1^{v-1}A_2 + A_1^{v-2}A_0 & A_1^{v-2}A_2 + A_1^{v-3}A_0 & \cdots & A_1A_2 + A_0 & A_2 \end{bmatrix},$$

$$\hat{L} = \begin{bmatrix} B & 0 & \cdots & 0 & 0 \\ A_1B & B & \cdots & 0 & 0 \\ \vdots & \vdots & & \vdots & \vdots \\ A_1^{v-1}B & A_1^{v-2}B & \cdots & A_1B & B \end{bmatrix}. \tag{8.10}$$

In this new 1D form, the dimension of state vectors is constant and consequently finding its regular form is possible. This sets the stage for designing specific 1D DSMC for the obtained 1D state space model (8.9), which is the subject of the next section.

8.3 DSMC for 1D Discrete Vector Form

Assume that $rank(\hat{L}) = v.m$ (matrix $\hat{L}$ is of full column rank), and the pair $(\hat{K},\hat{L})$ is controllable [27], [67].

Remark 8.4 *Since J is invertible, it is clear that the control matrix $\hat{L}$ in* (8.9) *is of full column rank if and only if the control matrix B in* (8.1) *is of full column rank.*

Since $rank(\hat{L}) = v.m$, there exists an orthogonal matrix $T_r \in \mathbb{R}^{[v.n]\times[v.n]}$ such that

$$T_r\hat{L} = \begin{bmatrix} 0_{[v.n-v.m]\times[v.m]} \\ \bar{L}_2 \end{bmatrix}, \tag{8.11}$$

where the matrix $\bar{L}_2 \in \mathbb{R}^{[v.m]\times[v.m]}$ and is nonsingular [98]. (Note that the orthogonal matrix T_r can be computed using QR decomposition [98]). After the coordinate transformation, we have

$$\begin{bmatrix} Z_1(i+1) \\ Z_2(i+1) \end{bmatrix} = \begin{bmatrix} \bar{K}_{11} & \bar{K}_{12} \\ \bar{K}_{21} & \bar{K}_{22} \end{bmatrix} \begin{bmatrix} Z_1(i) \\ Z_2(i) \end{bmatrix} + \begin{bmatrix} 0_{[v.n-v.m]\times[v.m]} \\ \bar{L}_2 \end{bmatrix} U(i) + T_r\hat{R}V(i), \tag{8.12}$$

where

$$\begin{bmatrix} Z_1(i) \\ Z_2(i) \end{bmatrix} = T_r X(i),$$

$$\begin{aligned} & Z_1(i) \in \mathbb{R}^{[v.n-v.m]} \text{ and } Z_2(i) \in \mathbb{R}^{v.m} \\ & \bar{K}_{11} \in \mathbb{R}^{[v.n-v.m]\times[v.n-v.m]} \\ & \bar{K}_{12} \in \mathbb{R}^{[v.n-v.m]\times[v.m]} \\ & \bar{K}_{21} \in \mathbb{R}^{[v.m]\times[v.n-v.m]} \\ & \bar{K}_{22} \in \mathbb{R}^{[v.m]\times[v.m]}. \end{aligned} \tag{8.13}$$

This representation is referred to as ' regular form' [98]. Furthermore, to design the sliding surface we ignore the term arising from the boundary conditions as it does not influence the stability. In these new coordinates the switching function becomes

$$\sigma_X(i) = \bar{S}_1 Z_1(i) + \bar{S}_2 Z_2(i), \tag{8.14}$$

where $\bar{S}_1 \in \mathbb{R}^{[v.m]\times[v.n-v.m]}$ and $\bar{S}_2 \in \mathbb{R}^{[v.m]\times[v.m]}$ satisfying $\bar{S} = [\bar{S}_1 \ \bar{S}_2] = ST_r^{-1}$. The design parameters $\bar{S}_1, \bar{S}_2$ determine the sliding surface and should be chosen such that, in the case that $\sigma_X(i) = 0$, all remaining dynamics are stable. During ideal sliding on the surface, $\sigma_X(i) = 0$ for all $k \geq k_s$, where k_s is the time when sliding starts, consequently

$$Z_2(i) = -\bar{S}_2^{-1}\bar{S}_1 Z_1(i). \tag{8.15}$$

Defining $\Omega = \bar{S}_2^{-1}\bar{S}_1$ and substituting the equation (8.15) into the equation (8.12) leads to:

$$Z_1(i+1) = (\bar{K}_{11} - \bar{K}_{12}\Omega) Z_1(i). \tag{8.16}$$

As a result, stability in the sliding mode is satisfied when all eigenvalues of the matrix $(\bar{K}_{11} - \bar{K}_{12}\Omega)$ are located inside the unit circle. Indeed, the problem of finding the matrix Ω is a classical state feedback problem. In [67], it is presented that if the pair $(\hat{K}, \hat{L})$ is controllable, the pair $(\bar{K}_{11}, \bar{K}_{12})$ is controllable as well. Therefore, any classical state feedback method can be used to compute Ω. Regarding the equation (8.15), the matrix $\bar{S}_2$ plays the role of a scaling parameter which can be selected arbitrarily but is invertible. For simplicity it is chosen as the following

$$\bar{S}_2 = \bar{L}_2^{-1}. \tag{8.17}$$

With the choice (8.17), the matrix Ω and consequently $\bar{S}_1$ can be found by pole placement, LQR-design or LMI methods. In this chapter the LQR-design method is used to find the matrix $\bar{S}_1$ by solving a certain discrete Riccati equation with proper choices of weighting matrices. In this case, the switching function can be obtained in the original coordinate as

$$S = \bar{S}_2[\Omega \quad I_{v.m}]T_r. \tag{8.18}$$

Now, to design a controller which guarantees the sliding mode of system the transformation matrix $T_s \in \mathbb{R}^{[v.n]\times[v.n]}$ is introduced as

$$T_s = \begin{bmatrix} I_{v.n-v.m} & 0_{[v.n-v.m]\times[v.m]} \\ \bar{S}_1 & \bar{S}_2 \end{bmatrix}. \tag{8.19}$$

This transformation matrix converts the system (8.12) to the following form:

$$\begin{bmatrix} Z_1(i+1) \\ \sigma_X(i+1) \end{bmatrix} = \begin{bmatrix} \tilde{K}_{11} & \tilde{K}_{12} \\ \tilde{K}_{21} & \tilde{K}_{22} \end{bmatrix} \begin{bmatrix} Z_1(i) \\ \sigma_X(i) \end{bmatrix} + \begin{bmatrix} 0_{[v.n-v.m]\times[v.m]} \\ I_{v.m} \end{bmatrix} U(i) + \begin{bmatrix} \tilde{V}_1 \\ \tilde{V}_2 \end{bmatrix}, \tag{8.20}$$

where $[\tilde{V}_1^T \quad \tilde{V}_2^T]^T = T_s T_r \hat{R} V(i)$ and $\tilde{V}_1 \in \mathbb{R}^{[v.n-v.m]}$, $\tilde{V}_2 \in \mathbb{R}^{v.m}$. In order to design a controller which forces the closed-loop system into the sliding mode we use the following linear reaching law presented in [95] and [138],

$$\sigma_X(i+1) = \Phi\sigma_X(i), \tag{8.21}$$

where the design parameter $\Phi \in \mathbb{R}^{[v.m\times v.m]}$ is chosen to be a diagonal matrix with all its diagonal elements ϕ_k, $k = 1,\ldots,v.m$, satisfying $0 \leq \phi_k < 1$.

Theorem 8.1 *Assuming the control input U as:*

$$U(i) = [\Phi - \tilde{K}_{22}]\sigma_X(i) - \tilde{K}_{21}Z_1(i) - \tilde{V}_2, \tag{8.22}$$

the system (8.20) *is stabilized.*

Proof 21 *Applying the above control law to the system* (8.20) *leads to the following closed-loop system:*

$$\begin{bmatrix} Z_1(i+1) \\ \sigma_X(i+1) \end{bmatrix} = \begin{bmatrix} \tilde{K}_{11} & \tilde{K}_{12} \\ 0 & \Phi \end{bmatrix} \begin{bmatrix} Z_1(i) \\ \sigma_X(i) \end{bmatrix} + \begin{bmatrix} \tilde{V}_1 \\ 0 \end{bmatrix}. \tag{8.23}$$

The poles of the closed-loop system are given by

$$\lambda(\mathscr{A}_{cl}) = \lambda(\tilde{K}_{11}) \cup \lambda(\Phi). \tag{8.24}$$

Obviously, the eigenvalues of Φ are assumed to be stable (by design choice). In addition, it can be easily proved that $\tilde{K}_{11} = \bar{K}_{11} - \bar{K}_{12}\Omega$ which is designed to be a stable matrix by (8.16). *Consequently, the system* (8.20) *is stabilized with control law* (8.22).

8.3.1 Direct method to find control law

It should be mentioned that instead of control law (8.22), another direct method is also possible to obtain the sliding control law [98]. Assuming that matrices $\bar{S}_1$ and $\bar{S}_2$ have been designed (by for instance LQR design) such that the reduced order dynamics (8.16) is stable. Now, by using linear reaching law (8.21) we have

$$\Phi\sigma_X(i) = SX(i+1). \tag{8.25}$$

Inserting equation (8.9) in (8.25) leads to

$$\Phi\sigma_X(i) = S[\hat{K}X(i) + \hat{L}U(i) + \hat{R}V(i)]. \tag{8.26}$$

Therefore, the control law can be defined to be

$$U(i) = (S\hat{L})^{-1}[\Phi SX(i) - S\hat{K}X(i) - S\hat{R}V(i)]. \tag{8.27}$$

This control law is called direct control law which can be obtained directly after computing the sliding matrix S.

8.4 Simulation Results

Consider the following 2D first FM model

$$A_1 = \begin{bmatrix} -0.56 & -0.33 \\ -0.10 & 0.45 \end{bmatrix}, A_2 = \begin{bmatrix} 0.33 & -0.54 \\ 1.26 & -0.41 \end{bmatrix},$$

$$A_0 = \begin{bmatrix} 0.51 & -0.09 \\ 0.00 & 0.04 \end{bmatrix}, B = \begin{bmatrix} 0 \\ 1 \end{bmatrix}. \tag{8.28}$$

Here $x \in \mathbb{R}^2$ and $u \in \mathbb{R}$. As seen, this 2D system with $U = 0$ is unstable. We assume this 2D system over the rectangle $\mu \times v$ ($\mu = 60$ and $v = 39$). Furthermore, it is supposed that

$$\begin{aligned} x(0,j) &= \begin{bmatrix} 1 \\ 1 \end{bmatrix}, \quad 0 \leq j \leq 39, \\ x(i,0) &= \begin{bmatrix} 0.2 \\ 0.2 \end{bmatrix}, \quad 0 \leq i \leq 60. \end{aligned} \tag{8.29}$$

To compute the orthogonal matrix T_r, Matlab QR command is used. According to (8.17), the matrix $\bar{S}_2 \in \mathbb{R}^{39\times 39}$ is chosen as $\bar{L}_2^{-1}$ (However, it is not necessary and can be chosen arbitrarily but invertible) and in addition matrix $\bar{S}_1 \in \mathbb{R}^{39\times 39}$ is determined by LQR design with state weighting of I_{39} and control weighting of $100I_{39}$. Eventually, matrix S is obtained from equation (8.18). $\Phi = 0.5I_{39}$ is used in control law (8.27) and the results of applying DSMC are given in Figures 8.1-8.3.

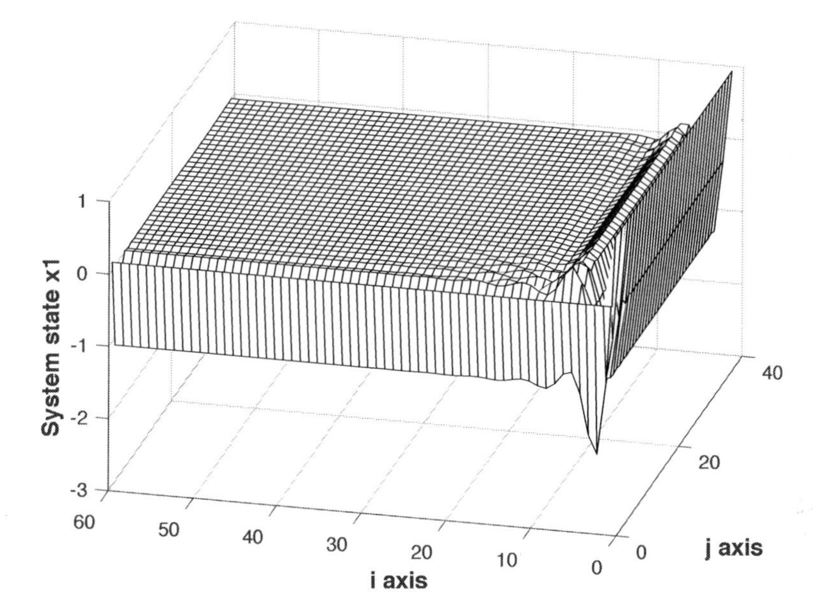

FIGURE 8.1
The system state x_1.

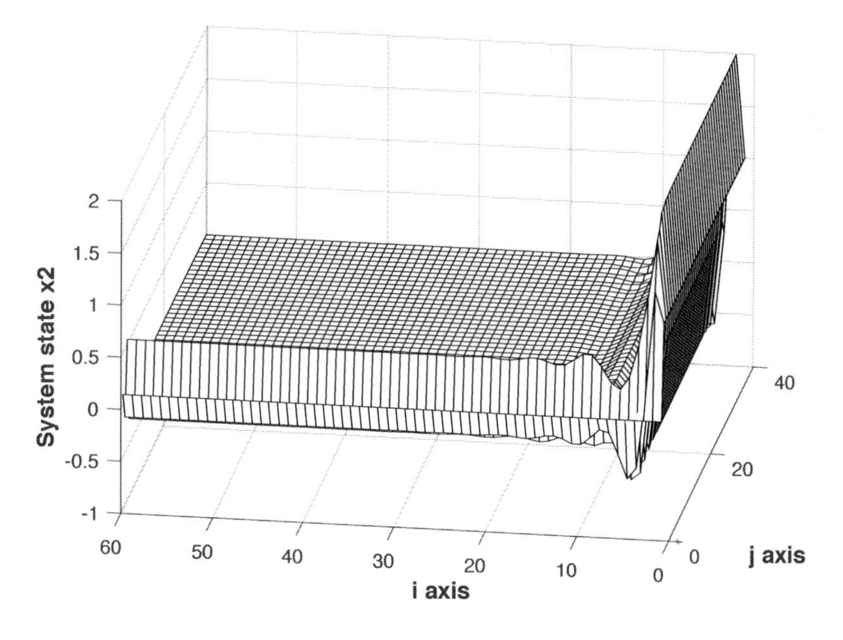

FIGURE 8.2
The system state x_2.

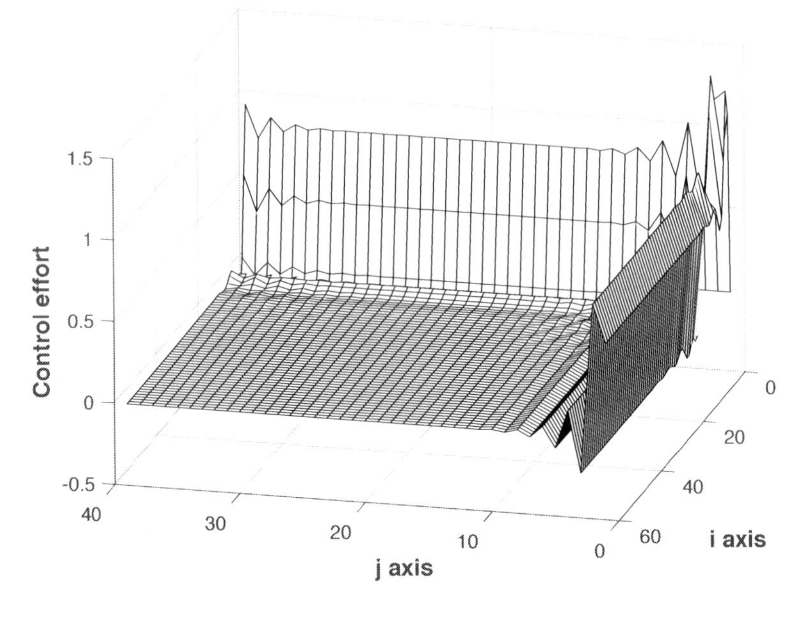

FIGURE 8.3
The control law u.

8.5 Conclusions

In this chapter, we have developed a new method to apply the DSMC to the 2D first FM model using the 1D vectorial form of 2D systems. Although the focus of this chapter has been on the first FM model, the derived results are more general and can be easily extended to other 2D models. In the proposed 1D vectorial form of this chapter, one of the 2D variables (i or j) is stacked and consequently the original 2D process is replaced by a 1D virtual process which can be controlled easily. Dealing with this new 1D form, the designing procedure of DSMC is more straightforward compared to the 2D system. Also, analyzing the controllability of the system is much easier in this form (which will be considered in the next chapter). Moreover, the proposed method of this chapter can be extended to the tracking problem in 2D systems.

9

Controllability analysis of 2D systems

CONTENTS

9.1 Introduction 171
9.2 WAM Model of the First FM Model 172
9.2.1 Controllability analysis of the WAM model 174
9.3 Controllability Analysis of the New Model in 8.2.1 178
9.3.1 Notion of local controllability for 2D systems 179
9.3.2 Directional controllability with respect to $\{j\}$-direction 180
9.3.3 Directional controllability with respect to $\{i\}$-direction 182
9.3.4 Directional minimum energy control input 183
9.4 Numerical Example 183
9.5 Conclusions 183

Abstract– Dealing with the one-dimensional (1D) form of two-dimensional (2D) systems is an alternative strategy to reduce the inherent complexity of 2D systems. To achieve the 1D form of 2D systems, a new row (column) process is considered in the previous chapter. The controllability analysis of this new 1D form is explored in this chapter. Two new notions of controllability named WAM-controllability and directional controllability for the underlying 2D systems will be defined. Corresponding conditions on the WAM-controllability and directional controllability will also be derived, which are particularly useful for the control problems of 2D systems via a 1D approach. Based on the presented directional controllability, a directional minimum energy control input is derived for 2D systems.

9.1 Introduction

Broadly speaking, the time domain analysis, such as controllability, reachability and observability of 2D systems, has been done by various researchers, resulting in a number of new notions such as local, global and causal controllability (reachability) [54, 53, 47, 46, 123, 48, 29, 28]. Necessary and sufficient conditions for the exact reconstructibility of the state of the second FM model have been presented in [28]. Other necessary and sufficient conditions with respect to 2D matrix polynomial equations for the local controllability and the causal reconstructibility of 2D linear systems are proposed in [123]. Reference [74] extends notions of local controllability, reachability, and reconstructibility for the general singular model of 2D linear systems.

In this chapter, first, the controllability analysis of the WAM model of the first FM model is studied, and a necessary condition for the controllability of this 1D model is given. It should be noted that finding the sufficient condition for the controllability of the WAM model is hard. This fact in addition to the time-varying form of the WAM model limits the applicability of the WAM model of 2D systems. This prompts us to exploit the row (column) process for converting 2D systems to their 1D models instead. On the other hand, during the procedure of designing the sliding surface in the previous chapter, it is assumed that the obtained 1D system is controllable; see e.g. [124] for the similar treatment. But, the controllability of the obtained 1D form and its relation to the original 2D system is an unanswered problem in Chapter 8. Hence, motivated by these issues, in this chapter, we focus on the controllability analysis of the proposed 1D form of the underlying 2D systems. Based on the controllability analysis, a new notion, *directional controllability*, for the underlying 2D systems is introduced and studied. More importantly, a necessary and sufficient condition for the directional controllability of 2D systems is presented in this chapter.

Also, there is a strong connection between controllability and the theory of minimal realization of linear time-invariant (LTI) control systems. Hence, the controllability result of this chapter would also provide useful insight into the observability and realization analysis of the underlying 2D systems using the developed 1D framework. Furthermore, note that the so-called minimum energy control problem is explicitly connected with controllability analysis [78]. Therefore, one application of the presented controllability analysis is the design of a specific 1D minimum energy control input for 2D systems called *directional minimum energy control input.* It should be noted that the results of this chapter are particularly useful for those who want to control the 2D systems via the proposed 1D framework.

The rest of this chapter is as follows. In the next section, the 1D WAM model of first FM systems and its controllability analysis are presented. Section 9.3 presents the controllability analysis of the 1D model presented in Chapter 8. Besides, a numerical example is given in Section 9.4. Finally, Section 9.5 concludes this chapter.

9.2 WAM Model of the First FM Model

Consider the first FM model in (8.1). In this section, a brief review on the WAM model [111] of the first FM model (8.1) is given. Then, the drawbacks of this method are explained, which motivate us to investigate an alternative 1D form for the 2D systems in the previous chapter which is more effective.

Define the state vectors $\phi(k)$ and $\nu(k)$ as

$$\begin{aligned}\phi(k) &= \begin{bmatrix}x^T(k,0) & x^T(k-1,1) & \cdots & x^T(0,k)\end{bmatrix}^T, \\ \nu(k) &= \begin{bmatrix}u^T(k,0) & u^T(k-1,1) & \cdots & u^T(0,k)\end{bmatrix}^T.\end{aligned} \tag{9.1}$$

The resulting WAM form of the first FM model (8.1) is as

$$\phi(k+1) = M(k)\phi(k) + N(k-1)\phi(k-1) + E(k-1)\nu(k-1). \tag{9.2}$$

Here, $M(k)$, $N(k-1)$ and $E(k-1)$ are determined by

$$\begin{aligned} M(k) &= \begin{bmatrix} I_{k+1} \\ 0_{1\times(k+1)} \end{bmatrix} \otimes A_2 + \begin{bmatrix} 0_{1\times(k+1)} \\ I_{k+1} \end{bmatrix} \otimes A_1, \\ N(k-1) &= T(k) \otimes A_0, \quad E(k-1) = T(k) \otimes B, \end{aligned} \tag{9.3}$$

where

$$T(k) = \begin{bmatrix} 0_{1\times k}^T & I_k & 0_{1\times k}^T \end{bmatrix}^T, \tag{9.4}$$

and I_k is the identity matrix of order k. Defining

$$r(k) = N(k-1)\phi(k-1) + E(k-1)\nu(k-1), \tag{9.5}$$

a 1D state space model is obtained as

$$\begin{bmatrix} \phi(k+1) \\ r(k+1) \end{bmatrix} = \begin{bmatrix} M(k) & I \\ N(k) & 0 \end{bmatrix} \begin{bmatrix} \phi(k) \\ r(k) \end{bmatrix} + \begin{bmatrix} 0 \\ E(k) \end{bmatrix} \nu(k). \tag{9.6}$$

Remark 9.1 *The state vector in* (9.6) *is a linear combination of the local states and inputs. However, in some applications, having state space equations with direct access to the local states is required. In this case, by introducing a new state vector,*

$$\begin{aligned}\hat{\phi}(k) = [&x^T(k,0), x^T(k,1), x^T(k-1,1), x^T(k-1,2), \\ &\cdots, x^T(1,k-1), x^T(1,k), x^T(0,k)]^T,\end{aligned} \tag{9.7}$$

a 1D state space equation with direct access to the state vectors $\phi(k)$ *and* $\phi(k+1)$ *is acquired.*

Remark 9.2 *In the definition of state vectors* (9.7)*, instead of using the local states just on the line* $i+j=k+1$*, the local states located on the line* $i+j=k$ *are also used to form state vectors. Generally, for the WAM description of 2D systems which are of at least second order, using the state vector* (9.7) *is useful. However, obtaining the WAM method for second order 2D systems (for instance the FM model) and especially for large scale 2D systems is complicated and, more importantly, the dimension of the state vector* (9.7) *is varying.*

Remark 9.3 *In the case that the boundary conditions are assumed to be constant, the state vector* (9.7) *should get rid of the boundary condition terms* $x(k,0)$ *and* $x(0,k)$ *as*

$$\begin{aligned}\bar{\phi}(k) = [&x^T(k,1), x^T(k-1,1), x^T(k-1,2), \\ &\cdots, x^T(1,k-1), x^T(1,k)]^T \in \mathbb{R}^{[(2k-1)\cdot n]}, \quad \forall k \geq 1.\end{aligned} \tag{9.8}$$

Hence, the 1D model is as follows

$$\bar{\phi}(k+1) = \bar{M}(k)\bar{\phi}(k) + \bar{E}(k)\nu(k) + \bar{V}(k), \tag{9.9}$$

where

$$\bar{M}(k) = \begin{bmatrix} A_2 & 0 & 0 & 0 & 0 & \cdots & 0 \\ I & 0 & 0 & 0 & 0 & \cdots & 0 \\ A_1 & A_0 & A_2 & 0 & 0 & \cdots & 0 \\ 0 & 0 & I & 0 & 0 & \cdots & 0 \\ 0 & 0 & A_1 & A_0 & A_2 & \cdots & 0 \\ \vdots & \vdots & \vdots & & & \cdots & \vdots \\ 0 & 0 & 0 & 0 & 0 & \cdots & I \\ 0 & 0 & 0 & 0 & 0 & \cdots & A_1 \end{bmatrix}, \tag{9.10}$$

$$\bar{E}(k) = \begin{bmatrix} B & 0 & 0 & \cdots & 0 \\ 0 & 0 & 0 & \cdots & 0 \\ 0 & B & 0 & \cdots & 0 \\ 0 & 0 & 0 & \cdots & 0 \\ \vdots & \vdots & \vdots & \cdots & \vdots \\ 0 & 0 & 0 & \cdots & B \end{bmatrix}, \bar{V}(k) = \begin{bmatrix} A_1x(k+1,0) + A_0x(k,0) \\ 0 \\ \vdots \\ 0 \\ A_2x(0,k+1) + A_0x(0,k) \end{bmatrix}, \quad \forall k \geq 1,$$

and, $\bar{M}(k) \in \mathbb{R}^{[(2k+1)\cdot n] \times [(2k-1)\cdot n]}$, $\bar{E}(k) \in \mathbb{R}^{[(2k+1)\cdot n] \times [(k+1)\cdot m]}$, $\bar{V}(k) \in \mathbb{R}^{(2k+1)\cdot n}$, for all $k \geq 1$. Note that, here, $\bar{M}(0) = A_2$, $\bar{E}(0) = B$, $\bar{V}(0) = A_1x(1,0) + A_0x(0,0)$ and $\bar{\phi}(0) = x(0,1)$.

9.2.1 Controllability analysis of the WAM model

This subsection aims to analyze the controllability of the 1D WAM model presented in (9.9). To this end, define the so-called state transition matrix $A^{i,j}$ as

$$\begin{aligned} A^{i,j} &= A_0A^{i-1,j-1} + A_1A^{i,j-1} + A_2A^{i-1,j} \\ &= A^{i-1,j-1}A_0 + A^{i,j-1}A_1 + A^{i-1,j}A_2, \quad \forall i,j > 0. \end{aligned} \tag{9.11}$$

Furthermore, it is assumed that

$$A^{0,0} = I_n,\ A^{-i,j} = A^{i,-j} = A^{-i,-j} = 0, \quad \forall i,j > 0. \tag{9.12}$$

Now, from (9.9) and with some recursive manipulations, we have

$$\bar{\phi}(k+1) - \hat{\mathscr{C}}_w(k)\mathscr{V}_w(k) - \prod_{i=0}^{k}\bar{M}(i)\bar{\phi}(0) = \mathscr{C}_w(k)\mathscr{U}_w(k), \tag{9.13}$$

where

$$
\begin{aligned}
\mathscr{C}_w(k) &= [\{\prod_{i=1}^{k}\bar{M}(i)\}\bar{E}(0) \mid \{\prod_{i=2}^{k}\bar{M}(i)\}\bar{E}(1) \mid \cdots \mid \bar{M}(k)\bar{E}(k-1) \mid \bar{E}(k)], \\
\hat{\mathscr{C}}_w(k) &= [\prod_{i=1}^{k}\bar{M}(i) \mid \prod_{i=2}^{k}\bar{M}(i) \mid \ \cdots \mid \bar{M}(k) \mid I_{\{(2k+1)\cdot n\}}], \\
\mathscr{U}_w(k) &= [v^T(0) \mid v^T(1) \mid \cdots \mid v^T(k-1) \mid v^T(k)], \\
\mathscr{V}_w(k) &= [\bar{V}^T(0) \mid \bar{V}^T(1) \mid \cdots \mid \bar{V}^T(k-1) \mid \bar{V}^T(k)].
\end{aligned}
\tag{9.14}
$$

As $\mathscr{V}_w(k)$ is determined by boundary conditions only (not a function of control), we neglect the second item of the equation (9.13) during the controllability analysis.

Theorem 9.1 *The 1D WAM model* (9.9) *is not controllable unless B is of full row rank.*

Proof 22 *Matrix $\mathscr{C}_w(k)$ in* (9.14) *can be found to be as*

$$
\mathscr{C}_w(k) = \left[\begin{array}{c|cc|c|cccc|ccccc}
A^{k,0}B & A^{k-1,0}B & 0 & \cdots & A^{1,0}B & \cdots & 0 & 0 & B & 0 & \cdots & 0 & 0 \\
A^{k-1,0}B & A^{k-2,0}B & 0 & \cdots & B & \cdots & 0 & 0 & 0 & 0 & \cdots & 0 & 0 \\
A^{k-1,1}B & A^{k-2,1}B & A^{k-1,0}B & \cdots & A^{0,1}B & \cdots & 0 & 0 & 0 & B & \cdots & 0 & 0 \\
\vdots & \vdots & \vdots & \vdots & \vdots & & \vdots & \vdots & \vdots & \vdots & & \vdots & \vdots \\
A^{1,k-1}B & A^{0,k-1}B & A^{1,k-2}B & \cdots & 0 & \cdots & A^{0,1}B & A^{1,0}B & 0 & 0 & \cdots & B & 0 \\
A^{0,k-1}B & 0 & A^{0,k-2}B & \cdots & 0 & \cdots & 0 & B & 0 & 0 & \cdots & 0 & 0 \\
A^{0,k}B & 0 & A^{0,k-1}B & \cdots & 0 & \cdots & 0 & A^{0,1}B & 0 & 0 & \cdots & 0 & B
\end{array}\right].
\tag{9.15}
$$

Left multiplying this matrix by

$$
\mathfrak{L}(k) = \begin{bmatrix}
I_n & -A_2 & 0 & \cdots & 0 \\
0 & I_n & 0 & \cdots & 0 \\
\vdots & \vdots & \vdots & & \vdots \\
0 & 0 & 0 & \cdots & I_n
\end{bmatrix},
\tag{9.16}
$$

where $\mathfrak{L}(k) \in \mathbb{R}^{[(2k+1)\cdot n]\times[(2k+1)\cdot n]}$, *it is obtained that*

$$\mathfrak{L}(k)\mathscr{C}_w(k) = \begin{bmatrix} 0 & 0 & \cdots & 0 & B & 0 & \cdots & 0 \\ A^{k-1,0}B & A^{k-2,0}B & \cdots & 0 & 0 & 0 & \cdots & 0 \\ * & * & \cdots & * & * & * & \cdots & * \\ \vdots & \vdots & \cdots & \vdots & \vdots & \vdots & \cdots & \vdots \\ * & * & \cdots & * & * & * & \cdots & * \end{bmatrix}. \tag{9.17}$$

where $\{*\}$ *means irrelevant entries. Note that since* $\mathfrak{L}(k)$ *is invertible, it does not change the row rank of the obtained matrix* $\mathfrak{L}(k)\mathscr{C}_w(k)$ *compared to* $\mathscr{C}_w(k)$*. Clearly, if B is not of full row rank, then,* $\mathfrak{L}(k)\mathscr{C}_w(k)$*, and consequently,* $\mathscr{C}_w(k)$ *is not of full row rank. Thus, the WAM model* (9.9) *is not controllable.*

From Theorem 9.1, it can be seen that the necessary condition for the controllability of the WAM model (9.9) is that B has full row rank. However, this condition is very restrictive. As mentioned in Remark 9.2, in order to construct the state vector $\bar{\phi}(k)$, the local states on the line $i+j=k+1$, and $i+j=k$ are both used. In other words, the even elements of the state vector $\bar{\phi}(\cdot)$ are carried elements from the previous step and only local states on the line $i+j=k+1$ have new information. Besides, the local states on the line $i+j=k+1$ will cover the whole space when k increases. As a result of this fact, the even block rows of the matrix $\mathscr{C}_w(k)$ are removed and the remaining matrix can be written as $\bar{\mathscr{C}_w}(k)$ as

$$\bar{\mathscr{C}_w}(k) = \left[\begin{array}{c|cc|c|cccc} A^{k,0}B & A^{k-1,0}B & 0 & \cdots & A^{1,0}B & \cdots & 0 & 0 \\ A^{k-1,1}B & A^{k-2,1}B & A^{k-1,0}B & \cdots & A^{0,1}B & \cdots & 0 & 0 \\ \vdots & \vdots & \vdots & \vdots & \vdots & & \vdots & \vdots \\ A^{1,k-1}B & A^{0,k-1}B & A^{1,k-2}B & \cdots & 0 & \cdots & A^{0,1}B & A^{1,0}B \\ A^{0,k}B & 0 & A^{0,k-1}B & \cdots & 0 & \cdots & 0 & A^{0,1}B \end{array}\right. \left|\begin{array}{cccccc} B & 0 & \cdots & 0 & 0 \\ 0 & B & \cdots & 0 & 0 \\ \vdots & \vdots & & \vdots & \vdots \\ 0 & 0 & \cdots & B & 0 \\ 0 & 0 & \cdots & 0 & B \end{array}\right], \tag{9.18}$$

which will be used to determine WAM- controllability defined below. This is equivalent to the output controllability with the following WAM output matrix for the system (9.9),

$$C_w(k) = \begin{bmatrix} I_n & 0 & 0 & \cdots & 0 \\ 0 & 0 & I_n & \cdots & 0 \\ \vdots & \vdots & & \cdots & \vdots \\ 0 & 0 & 0 & \cdots & I_n \end{bmatrix} \tag{9.19}$$

where $C_w(k) \in \mathbb{R}^{[k\cdot n]\times[(2k-1)\cdot n]}$. This equivalence to the special output controllability is in particular useful to the control/tracking problems for the 1D WAM model of the form (9.9), however it is beyond the scope of this chapter.

Definition 9.1 *The 2D system in* (2.1) *is said to be WAM-controllable if there exists a* $k \geq k^+ = \lceil \frac{2n}{m} - 2 \rceil \triangleq \min\{k \in \mathbb{N} \mid k \geq \frac{2n}{m} - 2\}$ *such that* $rank(\bar{\mathscr{C}_w}(k) \cdot \bar{\mathscr{C}_w^T}(k)) = (k+1) \cdot n$.

Remark 9.4 *It can be seen that* $\bar{\mathscr{C}_w}(k) \in \mathbb{R}^{[(k+1)\cdot n] \times [\frac{(k+1)(k+2)}{2} \cdot m]}$. *Besides, in the above definition, the condition* $k \geq \frac{2n}{m} - 2$ *is arising from the fact that the number of columns of matrix* $\bar{\mathscr{C}_w}(k)$ *is greater than or equal to the number of its rows if* $k \geq \frac{2n}{m} - 2$.

As $\bar{\mathscr{C}_w}(k)$ and its dimension are time-varying, one may ask about the future step's WAM- controllability even if the system (8.1) is WAM-controllable at the step k. Proposition 9.1, in the following, confirms the WAM- controllability for all the future steps, thus, validating the definition of WAM- controllability in Definition 9.1. Before it, consider the following lemma which provides a necessary condition for WAM-controllability.

Lemma 9.1 *If the system* (8.1) *is WAM-controllable, then the pairs* (A_1, *B*) *and* (A_2, *B*) *are both controllable.*

Proof 23 *It is obvious that the nonzero blocks of the first and the last block rows of the matrix* $\bar{\mathscr{C}_w}(k)$ *are equivalent to the* $(k+1)$*-th step controllability matrices of* (A_2, *B*) *and* (A_1, *B*)*, respectively. If either one is not controllable* $\bar{\mathscr{C}_w}(k)$ *is not of full row rank. Hence, system* (8.1) *is not WAM-controllable.*

Proposition 9.1 *If* $\bar{\mathscr{C}_w}(k)$ *is of full row rank for any* $k \geq k^+ = \lceil \frac{2n}{m} - 2 \rceil$, $\bar{\mathscr{C}_w}(k_1)$ *is of full row rank for any* $k_1 > k$.

Proof 24 *The matrix* $\bar{\mathscr{C}_w}(k+1)$ *can be rearranged by some column permutation operations (without changing the row rank) as*

$$\left[\begin{array}{cccccc|c} A^{k+1,0}B & A^{k,0}B & \cdots & A^{1,0}B & B & & 0 \\ \hline A^{k,1}B & A^{k-1,1}B & \cdots & A^{0,1}B & 0 & & \\ \vdots & \vdots & & \vdots & \vdots & & \bar{\mathscr{C}_w}(k) \\ A^{1,k}B & A^{0,k}B & \cdots & 0 & 0 & & \\ A^{0,k+1}B & 0 & \cdots & 0 & 0 & & \end{array}\right]. \tag{9.20}$$

From (9.20)*, it can be seen that if* $\bar{\mathscr{C}_w}(k)$ *has full row rank, necessarily, the controllability matrix of the pair* (A_2, *B*) *is of full row rank in* $(k+1)$*-th step from Lemma 9.1. Therefore, this pair is controllable in* $(k+2)$*-th step as well. Since the non-zero elements of the first block row of* (9.20) *contains the controllability matrix of the pair* (A_2, *B*) *in* $(k+2)$*-th step,* $\bar{\mathscr{C}_w}(k+1)$ *is of full row rank. This can be simply extended to the general case of* $\bar{\mathscr{C}_w}(k+r)$, $r \geq 1$.

The next result characterizes the WAM- controllability condition in terms of the original system matrices; if in particular $n = 2$, $m = 1$ and thus $\bar{\mathscr{C}_w}(2) \in \mathbb{R}^{6\times6}$, we would conclude the necessary and sufficient condition on the full rank of $\bar{\mathscr{C}_w}(2)$.

Theorem 9.2 *If $n = 2$, $m = 1$, the matrix $\bar{\mathscr{C}}_w(2)$ is of full row rank if and only if the three pairs (A_1, B), (A_2, B) and (A_0, B) are controllable.*

$$\bar{\mathscr{C}}_w(2) = \left[\begin{array}{c|cc|ccc} A_2^2 B & A_2B & 0 & B & 0 & 0 \\ (A_1A_2 + A_2A_1 + A_0)B & A_1B & A_2B & 0 & B & 0 \\ A_1^2 B & 0 & A_1B & 0 & 0 & B \end{array}\right]. \tag{9.21}$$

Proof 25 *Let $A_3 = A_1 + A_2$ and α_i, β_i be the scalar coefficients of the characteristic polynomial of A_i, satisfying $A_i^2 + \alpha_i A_i + \beta_i I = 0$, for $i = 1,2,3$. By noting $trace(A_i) = -\alpha_i$, it follows $\alpha_3 = \alpha_1 + \alpha_2$ for $n = 2$. Then*

$$\begin{aligned} A_1A_2 + A_2A_1 &= (A_1 + A_2)^2 - A_1^2 - A_2^2 \\ &= -\alpha_3(A_1 + A_2) - \beta_3 I + \alpha_1 A_1 + \beta_1 I + \alpha_2 A_2 + \beta_2 I \\ &= -\alpha_2 A_1 - \alpha_1 A_2 + (\beta_1 + \beta_2 - \beta_3)I. \end{aligned}$$

As a result, the matrix in (9.21) can be rewritten as

$$\bar{\mathscr{C}}_w(2) = \left[\begin{array}{c|cc|ccc} -(\alpha_2 A_2 + \beta_2 I)B & A_2B & 0 & B & 0 & 0 \\ \{-\alpha_2 A_1 - \alpha_1 A_2 + (\beta_1 + \beta_2 - \beta_3)I + A_0\}B & A_1B & A_2B & 0 & B & 0 \\ -(\alpha_1 A_1 + \beta_1 I)B & 0 & A_1B & 0 & 0 & B \end{array}\right].$$

With some elementary column operations on $\bar{\mathscr{C}}_w(2)$, $col_1 + \alpha_2 col_2 + \alpha_1 col_3 + \beta_2 col_4 + \beta_1 col_6 + (\beta_1 + \beta_2 - \beta_3)col_5$, ($col_i$ is the i-th block column of the matrix in (2.49)), one can change $\bar{\mathscr{C}}_w(2)$ to

$$\left[\begin{array}{c|cc|ccc} 0 & A_2B & 0 & B & 0 & 0 \\ A_0B & A_1B & A_2B & 0 & B & 0 \\ 0 & 0 & A_1B & 0 & 0 & B \end{array}\right]. \tag{9.22}$$

With some row and column permutations (9.22) is converted to

$$\left[\begin{array}{cc|cc|cc} A_0B & B & A_1B & 0 & A_2B & 0 \\ \hline 0 & 0 & A_2B & B & 0 & 0 \\ \hline 0 & 0 & 0 & 0 & A_1B & B \end{array}\right]. \tag{9.23}$$

It can be realized that all the rows of the above matrix are linearly independent if and only if the pairs (A_1, B), (A_2, B) and (A_0, B) are controllable. Note that, here, we use the fact that $\bar{\mathscr{C}}_w(2)$ is a square matrix and the row rank is equivalent to the column rank.

Theorem 9.2 provides a necessary and sufficient condition for the WAM- controllability of the special case $n = 2$, $m = 1$. As for the general case, while Lemma 9.1 presents the necessary condition for the controllability of the WAM model, finding its sufficient condition is hard and this can be the subject of future work. It will be shown, in Section 9.3, that the necessary and sufficient condition for the controllability of this 1D model, referred to as *directional controllability*, is only the controllability of one of the two pairs (A_1, B) and (A_2, B). The pair (A_0, B) would have no influence on the directional controllability.

9.3 Controllability Analysis of the New Model in 8.2.1

In this section, different controllability notions of the 2D system (8.1) are considered, namely, *local controllability* and *directional controllability*. This is achieved by studying the relation between the controllability of the obtained 1D system (8.4) and that of the 2D system (8.1).

9.3.1 Notion of local controllability for 2D systems

Different definitions of controllability according to different types of dynamical systems can be found in the literature. Broadly speaking, considering the controllability of 2D systems is relatively more complex compared to 1D systems. Instead of the notion of controllability introduced for 1D discrete-time systems, the notion of local controllability (reachability) is developed for 2D systems [79]. Here, the controllability of the first FM model (8.1) is studied referring to [47] and [79].

With the boundary conditions (8.6) and the given admissible controls sequence, it can be shown that

$$\begin{aligned} x(i,j) =& A^{i-1,j-1}A_0x(0,0) \\ &+\sum_{p=1}^{i}(A^{i-p,j-1}A_1+A^{i-p-1,j-1}A_0)x(p,0) \\ &+\sum_{q=1}^{j}(A^{i-1,j-q}A_2+A^{i-1,j-q-1}A_0)x(0,q) \\ &+\sum_{p=0}^{i-1}\sum_{q=0}^{j-1}A^{i-p-1,j-q-1}Bu(p,q), \end{aligned} \tag{9.24}$$

where the state transition matrix $A^{i,j}$ is as in (9.11) and (9.12). From (9.24), we have

$$\begin{aligned} M(i,j) \triangleq& x(i,j)-A^{i-1,j-1}A_0x(0,0) \\ &-\sum_{p=1}^{i}(A^{i-p,j-1}A_1+A^{i-p-1,j-1}A_0)x(p,0) \\ &-\sum_{q=1}^{j}(A^{i-1,j-q}A_2+A^{i-1,j-q-1}A_0)x(0,q) \\ =&\sum_{p=0}^{i-1}\sum_{q=0}^{j-1}A^{i-p-1,j-q-1}Bu(p,q) \\ =&\mathfrak{C}_{ij}u_{ij}, \end{aligned} \tag{9.25}$$

where

$$\mathfrak{C}_{ij} = [A^{i-1,j-1}B, A^{i-1,j-2}B, \cdots, A^{i-1,0}B, \cdots, A^{0,j-1}B, A^{0,j-2}B, \cdots, B], \tag{9.26}$$

and

$$\begin{aligned} u_{ij} = [& u^T(0,0), u^T(0,1), \cdots, u^T(0,j-1), \cdots, \\ & u^T(i-1,0), u^T(i-1,1), \cdots, u^T(i-1,j-1)]^T. \end{aligned} \tag{9.27}$$

Definition 9.2 *Consider the system* (8.1) *with the boundary condition* (8.6). *This system is locally controllable in a given rectangle* $[(0,0),(\mu,v)]$ *if for every boundary conditions* (8.6) *and for every vector* $x_d \in \mathbb{R}^n$, *there exists a sequence of controls* $u_{\mu v}$ *as in* (9.27) *such that* $x(\mu,v) = x_d$.

The matrix $\mathfrak{C}_{ij}$ in (9.26) is known as a local controllability matrix.

Lemma 9.2 ([79]) *The system* (2.1) *is locally controllable in a given rectangle* $[(0,0),(\mu,v)]$ *with unconstrained control inputs* u *if and only if rank*$(\mathfrak{C}_{\mu v} \cdot \mathfrak{C}_{\mu v}^T) = n$.

Furthermore, it is shown in [47] that Lemma 9.2 can be confined to the following lemma.

Lemma 9.3 *The system* (2.1) *is locally controllable in a given rectangle* $[(0,0),(\mu,v)]$ *with unconstrained control inputs* u *if and only if rank*$(\mathfrak{C}_{nn} \cdot \mathfrak{C}_{nn}^T) = n$ *where* $\mu \geq n$ *and* $v \geq n$.

It should be mentioned that this lemma is proven in [47] for a reachability case. Indeed, a method similar to proving the Cayley-Hamilton theorem for 1D systems can be developed for the 2D case.

9.3.2 Directional controllability with respect to $\{j\}$-direction

In this subsection, the controllability of the 1D system in (2.9) is considered. Moreover, a new notion of controllability for this special form of the 2D system in (8.1) is defined.

Now, define

$$\mathscr{M}(i) = [M^T(i,1) \cdots M^T(i,v)]^T = [(\mathfrak{C}_{i1}u_{i1})^T \cdots (\mathfrak{C}_{iv}u_{iv})^T]^T. \tag{9.28}$$

Since $u_{i1}, \ldots, u_{i(v-1)}$ are included in u_{iv}, (9.28) can be rewritten as

$$\mathscr{M}(i) = \mathscr{C}_i u_{iv}, \tag{9.29}$$

where $\mathscr{C}_i$ is

$$\mathscr{C}_i = \left[\begin{array}{ccccc|c} A^{i-1,0}B & 0 & \cdots & 0 & 0 & \cdots \\ A^{i-1,1}B & A^{i-1,0}B & \cdots & 0 & 0 & \cdots \\ \vdots & \vdots & & \vdots & \vdots & \vdots \\ A^{i-1,v-2}B & A^{i-1,v-3}B & \cdots & A^{i-1,0}B & 0 & \cdots \\ A^{i-1,v-1}B & A^{i-1,v-2}B & \cdots & A^{i-1,1}B & A^{i-1,0}B & \cdots \end{array}\right.$$
$$\left.\begin{array}{|ccccc} B & 0 & \cdots & 0 & 0 \\ A^{0,1}B & B & \cdots & 0 & 0 \\ \vdots & \vdots & & \vdots & \vdots \\ A^{0,v-2}B & A^{0,v-3}B & \cdots & B & 0 \\ A^{0,v-1}B & A^{0,v-2}B & \cdots & A^{0,1}B & B \end{array}\right]. \tag{9.30}$$

Lemma 9.4 *The matrix $\mathscr{C}_i$ in (9.30) satisfies*

$$\mathscr{C}_i = \left[\hat{K}^{i-1}\hat{L} \mid \cdots \mid \hat{K}\hat{L} \mid \hat{L}\right], \tag{9.31}$$

where $\hat{K}$ and $\hat{L}$ have the form in (8.10).

Proof 26 *From (8.9) it can be demonstrated that*

$$X(i) - \hat{K}^i X(0) - \mathscr{C}_{v_i}\mathscr{V}(i) = \mathscr{C}_{u_i}\mathscr{U}(i), \tag{9.32}$$

where

$$\begin{aligned} \mathscr{V}(i) &= [V^T(0)\cdots V^T(i-1)]^T, \mathscr{U}(i) = [U^T(0)\cdots U^T(i-1)]^T, \\ \mathscr{C}_{u_i} &= \left[\hat{K}^{i-1}\hat{L} \mid \cdots \mid \hat{K}\hat{L} \mid \hat{L}\right], \mathscr{C}_{v_i} = \left[\hat{K}^{i-1}\hat{R} \mid \cdots \mid \hat{K}\hat{R} \mid \hat{R}\right], \end{aligned} \tag{9.33}$$

and $U(\cdot), V(\cdot)$ are defined in (8.3). Noting that $\mathscr{U}(i) = u_{iv}$ and comparing (9.32) with (9.29), we can conclude (9.31) as $\mathscr{C}_i = \mathscr{C}_{u_i}$.

As $X(0)$ and $\mathscr{V}(i)$ are determined by the boundary and initial conditions, we only need to check $\mathscr{C}_i$ to analyze the controllability of system (9.32). As seen, the matrix $\mathscr{C}_i$ has the form of the controllability matrix of the 1D system (8.9), hence, the controllability of the 1D system (8.9) can be analyzed by checking the rank of this matrix. Furthermore, in the sequel it is shown that the matrix $\mathscr{C}_i$ in (9.30) has more to do with the local controllability of the 2D system in (8.1). Note that, in the sequel of this chapter, it is assumed that $\mu \geq n$ and $v \geq n$, without loss of generality.

Lemma 9.5 *The system (8.9) is controllable at the k-th ($k = 1, \cdots, \mu$) step with unconstrained control inputs U, if and only if rank$(\mathscr{C}_k \cdot \mathscr{C}_k^T) = v \cdot n$.*

Proof 27 *From Lemma 9.4, the k-th step controllability matrix of (8.9) is equivalent to $\mathscr{C}_k$. Hence, this system is controllable if and only if $\mathscr{C}_k$ has full row rank.*

Moreover, in the following theorem it will be shown that when $\mu \geq n$, $v \geq n$ and $\mathscr{C}_\mu$ is of full row rank, the local controllability matrix $\mathfrak{C}_{nn}$, and hence, $\mathfrak{C}_{\mu v}$ will be of full row rank. However, the converse of this issue is not always true.

Theorem 9.3 *The local controllability matrix $\mathfrak{C}_{nn}$ has full row rank if the matrix $\mathscr{C}_\mu$ has full row rank where $\mu \geq n$ and $v \geq n$.*

Proof 28 *$\mathscr{C}_\mu$ has v block rows with each block having the dimension $\{n \times (\mu \cdot v \cdot m)\}$. It is not hard to show that the nonzero blocks of the n-th block row of $\mathscr{C}_\mu$ is equivalent to the controllability matrix $\mathfrak{C}_{\mu n}$. Hence, if $\mathscr{C}_\mu$ has full row rank, $\mathfrak{C}_{\mu n}$ and thus $\mathfrak{C}_{\mu v}$ has full row rank. From Lemma 9.2, the 2D system (8.1) is locally controllable in a given rectangle $[(0,0),(\mu,v)]$. According to Lemma 9.3, it can be concluded that $\mathfrak{C}_{nn}$ is of full row rank.*

In other words, whenever the matrix $\mathscr{C}_\mu$ has full row rank the 1D form system (8.9) is controllable and the 2D system (8.1) is locally controllable in a given rectangle $[(0,0),(\mu,v)]$ with unconstrained control inputs.

Now comes the main result of this section.

Theorem 9.4 *The 1D form* (8.9) *of the 2D system* (8.1) *is controllable if and only if the matrix pair* (A_2, B) *is controllable.*

Proof 29 *By some column permutations (without changing the row rank) the matrix* $\mathscr{C}_n$ *is rearranged to*

$$\left[\begin{array}{ccc|ccc|c|ccc}
A^{n-1,0}B & \cdots & B & 0 & \cdots & 0 & \cdots & 0 & \cdots & 0\\
A^{n-1,1}B & \cdots & A^{0,1}B & A^{n-1,0}B & \cdots & B & \cdots & 0 & \cdots & 0\\
\vdots & & \vdots & \vdots & & \vdots & \vdots & \vdots & & \vdots\\
A^{n-1,v-2}B & \cdots & A^{0,v-2}B & A^{n-1,v-3}B & \cdots & A^{0,v-3}B & \cdots & 0 & \cdots & 0\\
A^{n-1,v-1}B & \cdots & A^{0,v-1}B & A^{n-1,v-2}B & \cdots & A^{0,v-2}B & \cdots & A^{n-1,0}B & \cdots & B
\end{array}\right]. \tag{9.34}$$

Obviously, the matrix in (9.34) *is a lower-triangular block matrix and its diagonal blocks are the controllability matrix of the pair* (A_2, B)*. Therefore, the controllability of* (A_2, B) *is equivalent to the controllability of* $(\hat{K}, \hat{L})$*.*

Here, according to Theorem 9.4, a new notion of controllability for 2D systems is defined.

Definition 9.3 *The 2D system in* (8.1) *is said to be directionally controllable with respect to the direction* $\{j\}$*, if its 1D form* Σ_v *in* (8.9) *is controllable.*

Proposition 9.2 *The 2D system in* (8.1) *is directionally controllable with respect to the direction* $\{j\}$*, if and only if the matrix pair* (A_2, B) *is controllable.*

Remark 9.5 *Basically, the notion of local controllability of 2D systems uses the Kalman- controllability notion and extends it to a more general form for 2D systems. Meantime, the notions of WAM controllability and/or directional controllability defined specifically for the 1D form of the 2D system* (8.1) *also exploit the standard Kalman- controllability notion. Note that Theorem 9.4 provides a sufficient and necessary condition for the controllability of the obtained 1D system* (8.9) *which is exactly equivalent to the Kalman- controllability of the matrix pair* (A_2, B)*.*

9.3.3 Directional controllability with respect to $\{i\}$-direction

In the procedure of [5] and this chapter, it is assumed that the $\{j\}$-direction is finite, and hence, the local states located in the same $\{j\}$-direction form the 1D stacking vectors. In the case that the $\{i\}$-direction is of finite dimension, the local states located in the same $\{i\}$-direction can be stacked to form the 1D stacking vectors. Similarly, a sufficient and necessary condition of the directional controllability with respect to $\{i\}$-direction can be obtained as follows.

Proposition 9.3 *The 2D system in* (2.1) *is directionally controllable with respect to the direction* $\{i\}$, *if and only if the matrix pair* (A_1, B) *is controllable.*

9.3.4 Directional minimum energy control input

For 1D LTI systems the controllability analysis is strongly related to the so-called minimum energy control problem [78]. In this subsection, a specific minimum energy control input is proposed for 2D systems according to the directional controllability notion given in the previous subsections. This control input will be denoted in this chapter as the *directional minimum energy control input.*

Suppose that (A_2, B) is controllable. From Theorem 9.4 we have that the matrix pair $(\hat{K}, \hat{L})$ of the system in (8.9) is controllable with $\mathscr{C}_{i_f}$, the controllability matrix, where $i_f > n$. Let

$$\tilde{\mathscr{U}}(i_f) = -\mathscr{C}_{i_f}^T(\mathscr{C}_{i_f}\mathscr{C}_{i_f}^T)^{-1}[X(i_f) - \hat{K}^{i_f}X(0) - \hat{\mathscr{C}}_{i_f}\mathscr{V}(i_f)], \tag{9.35}$$

($\tilde{\mathscr{U}}_{i_f}$ is as in (9.33)), then $\tilde{\mathscr{U}}(i_f)$ has the minimum energy $\left\|\tilde{\mathscr{U}}(i_f)\right\|^2$ among all possible control input sequences which can steer the system state from $X(0)$ to $X(i_f)$ [59]. The control input sequence given in (9.35) is said to have the directional minimum energy with respect to $\{j\}$-direction.

9.4 Numerical Example

Consider the following 2D first FM model

$$A_1 = \begin{bmatrix} -0.56 & 0 \\ 0 & 0 \end{bmatrix}, \; A_2 = \begin{bmatrix} -0.33 & -0.54 \\ 0.26 & -0.41 \end{bmatrix},$$
$$A_0 = \begin{bmatrix} 0.51 & -0.09 \\ 0.00 & 0.04 \end{bmatrix}, \; B = \begin{bmatrix} 0 \\ 2 \end{bmatrix}. \tag{9.36}$$

Here, $x \in \mathbb{R}^2$ and $u \in \mathbb{R}$. We assume this 2D system over the rectangle $\mu \times \nu$ ($\mu = 20$ and $\nu = 5$). It is supposed that $x(0, j) = \left[\begin{smallmatrix} 0 \\ 0 \end{smallmatrix}\right]$, $0 \le j \le 5$, $x(i, 0) = \left[\begin{smallmatrix} 0 \\ 0 \end{smallmatrix}\right]$, $0 \le i \le 20$. It can be seen that despite the uncontrollability of the pair (A_1, B), and, since the pair (A_2, B) is controllable, the pair $(\hat{K}, \hat{L})$ is controllable. Also, $\mathscr{C}_2$ and $\mathfrak{C}_{22}$ have full row rank ($rank(\mathscr{C}_2) = 6$ and $rank(\mathfrak{C}_{22}) = 2$). As a result, this 2D system can be said to be *directionally controllable with respect to the* $\{j\}$-direction. Note that since $rank(\bar{\mathscr{C}}_w(2)) = 5 < 6$ this system is not WAM controllable in the $3rd$ step. Also, since (A_1, B) is not controllable, from Lemma 9.1, this system is not WAM controllable in general.

Now, the results of applying the open-loop minimum energy control input sequences in (9.35), with $i_f = \mu = 20$, $X(0) = \mathbf{0}_{10\times 1}$ and $X(20) = 10 \times \mathbf{1}_{10\times 1}$, to the system (8.9) are shown in Fig. 9.1.

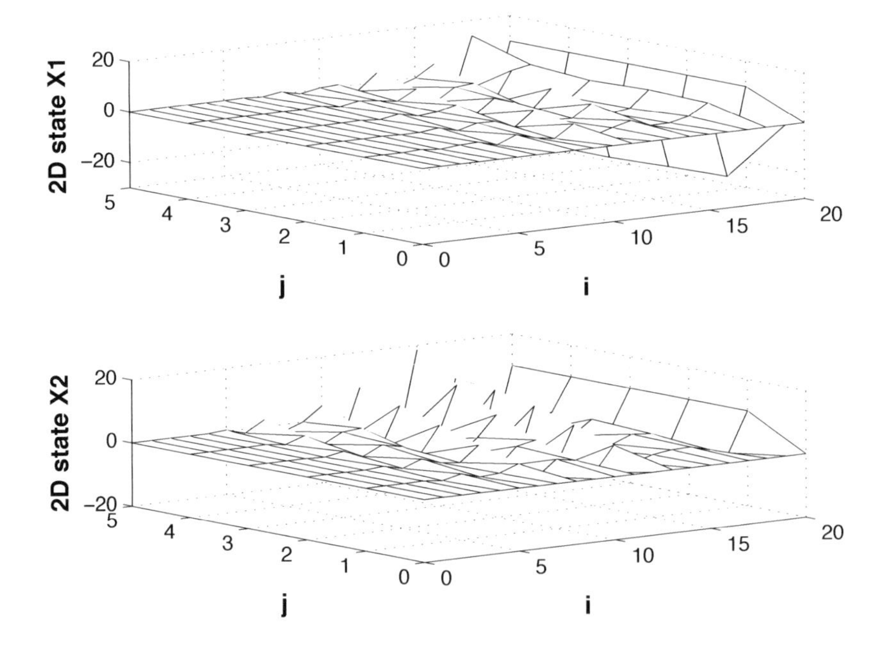

FIGURE 9.1
2D system state

9.5 Conclusions

In this chapter, a new WAM- controllability notion has been defined for 2D systems and a necessary condition is given accordingly. Then, a necessary and sufficient condition has been derived for the controllability of the newly proposed 1D model in Chapter 8. Accordingly a new notion, directional controllability, has been defined for the underlying 2D systems. The directional controllability analysis presented in this work is beneficial in terms of designing the so-called minimum energy control input.

Part V

Integral DSMC for Heart Rate Regulation

10

Heart rate regulation during cycle-ergometer exercise via event-driven biofeedback

CONTENTS

10.1 Introduction 190
10.2 Methods 193
  10.2.1 Equipment and data acquisition system 193
  10.2.2 HR profile 194
  10.2.3 Control system 195
    10.2.3.1 Actuator-based event-driven PID controller 195
    10.2.3.2 Actuator-based event-driven adaptive ISMC 196
    10.2.3.3 Auditory converter 198
    10.2.3.4 The proposed control mechanism in summary 199
    10.2.3.5 Two novel anti-windup mechanisms 200
    10.2.3.6 Relay controller strategy 201
  10.2.4 Tuning the PID controller gains 202
10.3 Results 202
10.4 Discussion 203
  10.4.1 Necessity of the project 205
  10.4.2 Discussion of the control system 206
  10.4.3 Discussion of the results 212
10.5 Conclusions 213

Abstract– This chapter investigates the problem of heart rate regulation during cycle-ergometer exercise using both a non-model-based and a model-based control strategy along with a real-time damped parameter estimation scheme. The model-based control strategy is a time-varying integral sliding mode controller. A recursive damped parameter estimation method is also developed, by incorporation of a weighting upon the one-step parameter variation, which in contrast to the conventional parameter estimation schemes can avoid the occurrence of the so-called blowup phenomena. Delivering a feedback signal when the pedals are not in a suitable position to efficiently exert force may be ineffective and this may, in turn, lead to the cognitive disengagement of the user from the feedback controller. Therefore, the calculated control signals are transmitted to the subjects employing a synchronized biofeedback mechanism. This chapter examines a novel form of control system which has been designed for this project. The system is called an "actuator-based event-driven control system". The proposed control and estimation scheme were experimentally verified using several healthy male participants and the results demonstrated that the designed scheme is able to regulate the HR of the exercising subjects to a predetermined HR profile.

10.1 Introduction

It is generally accepted that regular aerobic exercise has a positive effect on reducing the risk of heart disease, rehabilitation post infarct and for reversing the health effects of diabetes or being overweight [36, 148, 22, 94].

Heart rate (HR) can be used as an index to monitor the exercise intensity [106]. This fact makes it simpler to develop a control system to steer the human HR to a predetermined and individual exercise prescription, represented as a target HR profile, instead of directly employing the exercise intensity in kJoules or Watts and exercise rate (ER) as control parameters.

A number of strategies have explored real-time control of HR during treadmill exercises such as, e.g. classical proportional, integral and derivative (PID) control [76], H_∞ control [34, 132] and model predictive control [129, 130]. However, to date, this issue has not received very much attention, for cycle-ergometer exercises. In the case of a treadmill, the controller directly controls the treadmill speed and/or treadmill gradient, and as a result, the human does not play the role of actuator in the control-loop. Furthermore, they only have to passively respond to the variation of the controlled parameters (speed and/or elevation) of a treadmill. This also happens for the controllable automatic braking cycle-ergometers, in which the controller controls the load on the cycling system by changing the resistance of the brake device [75]. The problem of heart rate regulation during cycle-ergometer exercises can be considered in an entirely different framework. In other words, the controller commands can be transmitted directly to the exercising subjects and, then the human operator acts in the role of the actuator of the control system. This is commonly referred to as biofeedback with the operator in the feedback loop [52]. This novel framework obviates the need for controllable automatic brake equipment in the cycle-ergometers. In [25], a periodic auditory tone is used to inform the subject of how hard they must work. In other words, a translating function, according to the level of the controller signal, simply changes the auditory rate and frequency to show whether more or less effort is required. However, the control system in [25] generates and transmits biofeedback without considering the position of the exercising subjects' feet. This makes the control system a simple time-driven one, routine in the control engineering, which implements the control calculations at a fixed rate. However, this may lead to a cognitive disengagement of the subject from the feedback controller. One possible alteration to this audio output method might be adjusting the time duration and the frequency of the auditory stimulus, while synchronizing the rate of transmitting the auditory stimulus to the user with the position of the exercise bike pedals to optimize effective application of force. Implementing this idea, however, requires a new control system strategy (see Fig. 10.1), regarded here as an *actuator-based event-driven control system*, which is different from the existing *sensor-based event-driven control system* in the literature [61].

Owing to the simplicity of the PID control structure, it represents a simple and

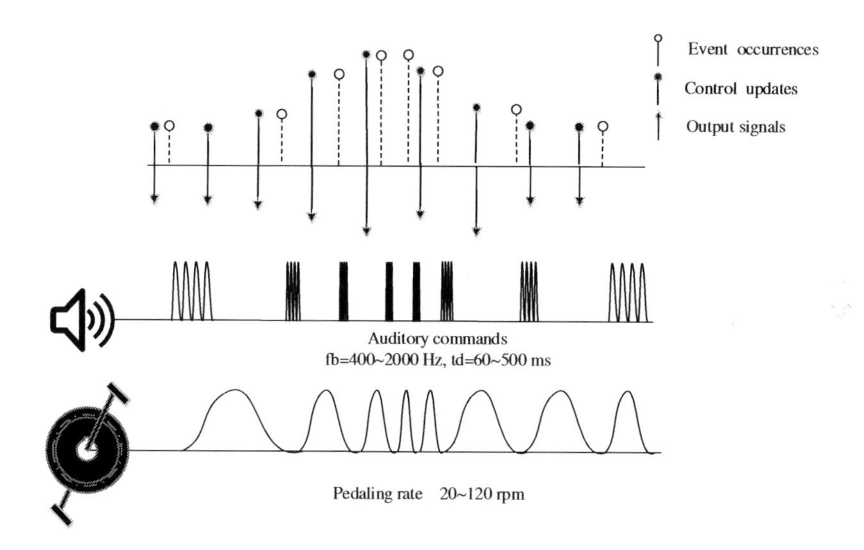

FIGURE 10.1
Mechanism of the proposed control system.

powerful control strategy which can be used for various industrial and practical applications. When using PID control, the controller parameters for each subject should be determined continuously using a subject specific model. Alternatively, one can design a robust PID controller which is able to compensate for the inter-individual differences in the dynamic HR response to the work rate [75]. The idea of a non-model-based control strategy for heart rate regulation during cycle-ergometer is depicted in Fig. 10.2.

On the other and, it is well known that the HR response to dynamic exercise is nonlinear and it may be different for each exercising subject, in different physical situations [25]. Hence, an adaptive and/or model-based control method is usually preferable to address the problem of HR control. In this work, we additionally developed an adaptive integral sliding mode control strategy, along with a damped online parameter estimation scheme. Traditional SMC approaches may lead to a deadbeat control which is undesirable in practical applications because of the possible required large control efforts. The integral sliding mode control (ISMC) design resolves this issue by removing the poles at zero [1]. On the other hand, the traditional parameter estimation schemes (e.g. recursive least squares (RLS) method) provide no useful

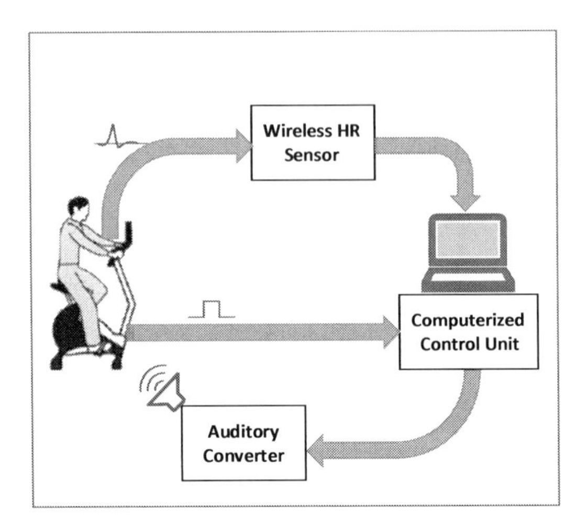

FIGURE 10.2
Block diagram for the HR regulation system during cycling exercise using non-model-based control schemes.

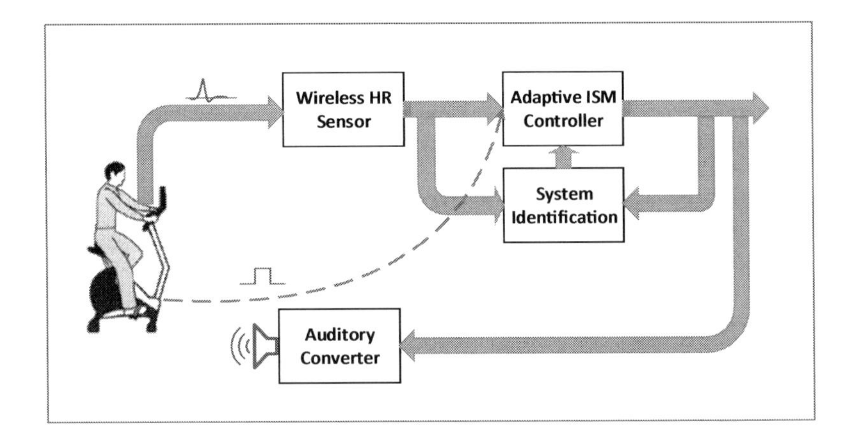

FIGURE 10.3
Block diagram for the HR regulation system during cycling exercise using model-based control schemes.

parameter estimates while the excitation signal is small or zero. This issue is referred to as *blowup* phenomena in the literature. Furthermore, if for a long period no excitation happens, large fluctuations may happen in the estimated parameters. The reference [84] modifies the conventional RLS method to bound the amount of parameter variations. Inspired by [84], this work developed a damped RLS scheme using a regularization approach that is specifically useful for HR response dynamics identification. Fig. 10.3 illustrates the idea of a model-based control strategy for heart rate regulation during cycle-ergometer.

In this chapter, an event-driven control system is proposed, in which two anti-windup [24] mechanisms are designed to protect the user against possible large HR fluctuations owing to inaccurate controller parameters and/or individual HR responses. This event-driven control system consists of a PID or an adaptive ISMC controller whose output signal is translated to a set of auditory stimulus transmitted to the exercising subject at a time-varying rate synchronized with the subject's pedaling rate and position of the pedals.

For comparison, we also used a simple switching (relay) controller to generate visual phrases which can be read by the exerciser, as well as a conventional PID controller whose output signal is transmitted as biofeedback (auditory signal), however, with a fixed transmission rate or equivalently without considering the position of the exercising subjects' feet.

The novel control system was experimentally verified employing exercising subjects. It is worth noting that statistical analysis presented in [16] revealed that the proposed actuator-based event-driven PID controller has better performance compared to a conventional PID controller with a fixed-rate biofeedback mechanism or a relay controller using synchronized visual biofeedback.

10.2 Methods

10.2.1 Equipment and data acquisition system

In this project, we utilized a *Nonin 4100 Pulse Oximeter* to collect the HR data during exercise; see Fig. 10.4. Various data formats can be extracted from the Nonin 4100 wireless sensor such as SpO_2, HR, Plethysmographic pulse data etc. Although this sensor outputs different data, which makes it an adequate choice for clinical applications, it may sometimes show some abnormalities in the data. This is because the Nonin 4100 pulse oximeter relies on comparing two different light waves and an algorithm to calculate the heart rate and oxygen saturation levels. If the sensor slips on the finger, it is unable to pick up a clean feed of data and therefore errors in the data are generated. Furthermore, if a subject clamp their fingers around an object while on the bike, blood may be constrained or cut off from the finger, resulting in a lack of data or incorrect data being collected. To resolve this we need to ensure that on the cycle-ergometer the subject's arms rested on the supporting structure

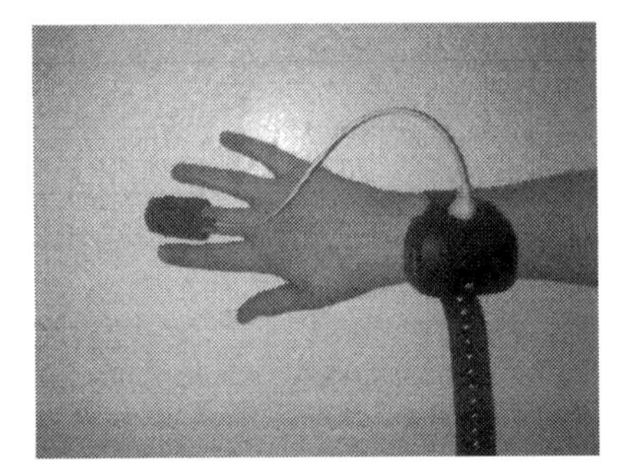

FIGURE 10.4
Nonin 4100 Pulse Oximeter

while the fingers remained comfortably stretched out to ensure an adequate flow of blood. Moreover, the cycle-ergometer that we used in this project is an air vane exercise bike, which is well suited for safe exercise by the frail and/or elderly. Since it has a big supporting structure, which can be used by the exercising patient to stabilize themselves during exercising, the finger pulse oximeter is almost completely stable during exercise and it generates minimal artefacts. We have developed a data acquisition system using *National Instrument LabVIEW* which provides easy synchronization and a graphical user interface. The LabVIEW collected HR signal from the Nonin 4100 pulse oximeter every 1 s. In addition, to prepare a reliable real-time HR signal for the computer controller, an exponentially weighed moving average filter (see [39]), with filter coefficient $\alpha = 0.75$, is implemented.

Two reed switches and magnets have been attached to the crank shaft of the pedals of the exercise bike, in order to provide a pulse whenever a full revolution has been completed; see Fig. 10.5. In the Labview program, a time-delay parameter may be set to help the user to adjust the time-delay between the position of the pedal sensed by the sensor and the time when the command is sent. This parameter is useful since we can adjust the point where the user can most effectively apply force to the pedals; see Fig. 10.5.

10.2.2 HR profile

The HR profile in this work is selected in three stages. In the first stage, called the *warm up* period, we aim to gradually increase the user HR from their normal HR (HR_n) to the *exercise* HR (HR_e) which then remains constant for a while (T_e). The warm up slope is

$$\theta_w = \frac{HR_e - HR_n}{T_w}, \tag{10.1}$$

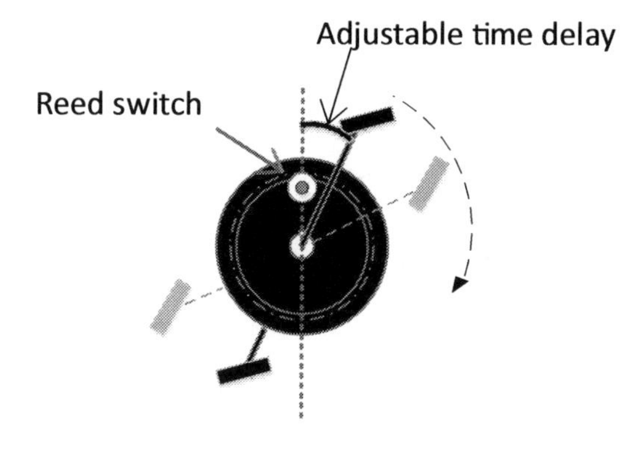

FIGURE 10.5
Reed switch and adjustable time delay parameter

in which θ_w represents the warm up slope and T_w shows the warm up time duration. The warm up time duration T_w is used to avoid an excessively rapid increase in the HR. The second aerobic training stage is denoted by T_e, which may be computed according to the *maximum* HR. For obtaining the ideal target aerobic training HR, various formulas have been proposed. The *maximum* HR (HR_{max}) is commonly characterized by the well-known *Haskell and Fox* formula [115] as:

$$HR_{max} = 220 - \text{Age}. \tag{10.2}$$

We then set the exercising HR_e as $0.65 \sim 0.85$ of HR_{max}, depending on the subject's level of fitness and/or risk profile. The application of the third *cooling down* or *recovery* stage (T_c) is basically to prevent possible venous pooling and reduced venous return to the heart from an excessively abrupt termination of the exercise.

Remark 10.1 *In this project, the HR profile is dynamic. In other words, the HR profile always starts from the subject's current resting HR and increases to a value which is calculated based on the age of the subject (aerobic training ideal HR).*

10.2.3 Control system

10.2.3.1 Actuator-based event-driven PID controller

Since this project does not use an additional automatic braking system on the cycle-ergometer, the use of biofeedback is necessary to control the exercise intensity. This means that the controller commands are sent directly to the exerciser and they must be interpreted and implemented by the exerciser as a part of the control loop. We use the information from two sensors installed on the pedaling system to trigger the biofeedback with a suitable delay to optimize the ability of the user to both interpret the command and to assert the appropriate amount of force on the pedals at the correct time. However, as the rate of output (HR) sampling is not synchronous with the rate of transmitting the controller commands (auditory signals) to the user, through the new proposed biofeedback mechanism, we needed to develop a novel control system which is different from the conventional control systems, in which the sampling rate of the system outputs is assumed to be the same as the rate of control signal transmission to the system actuators. This section briefly explains the novel control system designed for this project. A discussion on the control system will be given later in this chapter.

Consider a time-driven PID controller in the velocity form such as

$$u(k) = u(k-1) + \Delta u(k), \tag{10.3}$$

where

$$\begin{aligned} \Delta u(k) =& K_p[e(k) - e(k-1)] + K_i T_s e(k) \\ &+ \frac{K_d}{T_s}[e(k) - 2e(k-1) + e(k-2)], \end{aligned} \tag{10.4}$$

in which

$$e(t) = r(t) - y(t), \tag{10.5}$$

where $u(t)$, $r(t)$, $y(t)$, K_p, K_i and K_d are control effort, reference, actual output, proportional gain, integral gain and derivative gain, respectively, and T_s is a fixed sampling rate and is equivalent to the sampling period of the HR sensor. Then, whenever an event occurs (the left or right pedal passes the reed sensor), the last prepared update of the control signal will be transmitted as a biofeedback signal. As seen, the sampling rate of the HR transmitted to the program by the HR transducer, and thus, the updating rate of the control signal in (10.3), are typically constant (1 sample per second). However, the rate of biofeedback transmission to the exercising subject varies based on the pedaling rate. This specific form of control system, called here the *actuator-based event-driven control system*, will be discussed later in Section 10.4.2.

10.2.3.2 Actuator-based event-driven adaptive ISMC

Model identification – Damped least squares (LS) method:

Since the HR response to dynamic exercise is time-varying, a stationary control scheme may not provide adequate performance. Therefore here an adaptive control scheme is utilized to deal with the time-varying dynamics.

The conventional least squares method obtains the parameter estimate minimizing the following quadratic cost function:

$$J_0(\hat{\theta}) = \sum_{t=k-N}^{k} \beta^{k-t}[y(t) - \phi(t-1)^T \hat{\theta}]^2, \tag{10.6}$$

where $y(k)$ is the process output, $\hat{\theta}$ is the parameter estimate, $\phi(k)$ is the regressor vector, N is the sample number, and β is the exponential forgetting factor. Now, the goal is to find model parameter estimates $(\hat{\theta})$ that minimize the least squares of the differences between the actual system outputs and the predicted values for the system outputs. The vector $\phi(k)$ consists of the past input and output values. Further, $0 < \beta \leq 1$, referred to as the forgetting factor, is a scalar that provides us with a flexibility to weight the influence of the past information on the estimation; see [23]. However, the traditional RLS method suffers from a drawback. For instance, if the excitation signal is small and all elements of $\phi(k)$ are zero, no useful parameter estimates can be achieved. This issue is called *blowup* in the literature. In simpler terms, using a constant forgetting factor, if for a long period no excitation happens, the estimator will lose the proper values of the parameters, and the regressor vector will be dominated by noise. In such a case, large fluctuations may happen in the estimated parameters. The reference [84] modifies the conventional RLS method to bound the amount of parameter variation relative to previous-time parameter vector. This can be formulated as:

$$J(\hat{\theta}) = \sum_{t=k-N}^{k} \beta^{k-t}[y(t)-\phi(t-1)^T\hat{\theta}]^2 + \lambda_d(k)\left\|\hat{\theta}(k) - \hat{\theta}(k-1)\right\|,$$

where $\lambda_d(k)$ is a time-varying regularization parameter that implies the emphasis on bounding the parameter increments; i.e. $\lambda_d(k) = \lambda_d(k-1) = 0$ converts the above objective function exactly to the standard recursive least squares with exponential forgetting, or if $\lambda_d(k) = \infty$ ($\lambda_d(k)$ is set to a very large value) then the estimated parameters are immediately frozen to their current values.
Minimization of the above objective function results in the recursive equations as:

$$\varepsilon(k) = y(k) - \phi(k-1)^T\hat{\theta}(k-1), \tag{10.7}$$

$$\begin{aligned}\hat{\theta}(k) =& \hat{\theta}(k-1) + P(k-1)\phi(k-1)\varepsilon(k) \\ &+ P(k-1)\beta\lambda_d(k-1)[\hat{\theta}(k-1) - \hat{\theta}(k-2)],\end{aligned} \tag{10.8}$$

$$P(k-1)^{-1} = \beta P(k-2)^{-1} + \phi(k-1)\phi(k-1)^T + [\lambda_d(k) - \beta\lambda_d(k-1)]I, \tag{10.9}$$

where $\hat{\theta}(0) = \theta_0$ (θ_0 is a nominal parameter vector) and $P(0) = \lambda_d(0)^{-1}I$.

Now, we aim to model the human HR response to the control inputs. The model input here is the control commands that the controller generates. The relationship between the input and output, by assuming first order dynamics for the HR response, is represented as:

$$y(k+1) = a(k)y(k) + b(k)u(k) + d(k), \tag{10.10}$$

where $y(k) \in \mathbb{R}$, $u(k) \in \mathbb{R}$ and $d(k) \in \mathbb{R}$ denote the HR, the control input, and the disturbance (which is assumed to be smooth and bounded), respectively. Further, $a(k)$ and $b(k)$ are the time-varying system parameters which are scalars.

Discrete-time Integral Sliding Mode Control:

Consider the first order system in (10.10). Now the integral sliding surface is proposed as

$$\sigma(k) = e(k) - e(0) + \varepsilon(k)$$
$$\varepsilon(k) = \varepsilon(k-1) + \mu e(k-1), \tag{10.11}$$

where $e(k) = r(k) - y(k)$ denotes the tracking error ($r(k)$ is the reference trajectory), $\sigma(k)$, $\varepsilon(k)$ are the sliding function and integral variable and μ is the design parameter. Now letting $\sigma(k+1) = 0$ (which is a routine in sliding mode control) leads to:

$$\begin{aligned}\sigma(k+1) &= e(k+1) + \varepsilon(k+1) - e(0) \\ &= e(k+1) - (1-\mu)e(k) + \varepsilon(k) + e(k) - e(0) \\ &= e(k+1) - (1-\mu)e(k) + \sigma(k) = 0.\end{aligned}$$

Hence utilizing (10.10), one may show that the so-called equivalent controller is

$$u_{eq}(k) = b(k)^{-1}[r(k+1) - \rho e(k) - a(k)y(k) - d(k) + \sigma(k)],$$

where $\rho = 1 - \mu$. However, it is not possible to implement the above controller as $d(k)$ is unknown. As this work uses a moving average filter for smoothing the HR signal, and further, the updating rate of the system model is very much slower than the sampling rate, the disturbance estimate can be employed in the controller as

$$u(k) = b(k)^{-1}[r(k+1) - \rho e(k) - a(k)y(k) - \hat{d}(k) + \sigma(k)], \tag{10.12}$$

where $\hat{d}(k)$ is the disturbance estimate which can be achieved by the following equation:

$$\hat{d}(k) = d(k-1) = y(k) - a(k-1)y(k-1) - b(k-1)u(k-1). \tag{10.13}$$

Stability analysis:

By applying the controller in (10.12) to the system (10.10), we obtain

$$y(k+1) = r(k+1) - \rho e(k) + d(k) - \hat{d}(k) + \sigma(k).$$

Since $\sigma(k) = \hat{d}(k-1) - d(k-1)$, then

$$e(k+1) = \rho e(k) + \kappa(k), \tag{10.14}$$

in which $\kappa(k) = d(k) - 2d(k-1) + d(k-2) = O(T^3)$ (see [1]). As $\rho = 1 - \mu$, one may choose μ such that $\rho < 1$. This suffices that the tracking error dynamics remain bounded as $\kappa(k)$ is bounded as well.

10.2.3.3 Auditory converter

In order to instruct the user to apply more or less effort at the right time, a combination of varying both the time duration and the frequency of auditory stimulus was used in this project. In this method a shorter time duration and lower frequency are used to motivate the subject to reduce the application of force to the pedals and conversely to increase the applied force if the time duration and frequency increases. However, the time intervals and frequency range which can be used are limited. We know that the time durations of more than a specific value result in the so-called aliasing phenomena. Additionally, time durations which are less than a specific value may not be heard by the subjects. This issue also holds for the frequency of the auditory signal. Accordingly, we selected the following range of time duration and frequency for the auditory stimulus: $60 \sim 500\ ms$ and $400 \sim 2000\ Hz$, respectively. The auditory converter thereby contains two saturation functions,

$$t_d = \text{sat}\{g(u(k))\} = \begin{cases} \lambda_1 & g(u(k)) \leq \lambda_1 \\ g(u(k)) & \lambda_1 < g(u(k)) < \lambda_2 \\ \lambda_2 & g(u(k)) \geq \lambda_2, \end{cases} \tag{10.15}$$

where t_d is the time duration of the auditory signals and $g(u(k))$ is a scaling function which converts the range of $u(k)$ to the range of the time duration, $\lambda_1 = 60\ ms$, $\lambda_2 = 500\ ms$, and

$$f_b = \text{sat}\{h(u(k))\} = \begin{cases} \gamma_1 & h(u(k)) \leq \gamma_1 \\ h(u(k)) & \gamma_1 < h(u(k)) < \gamma_2 \\ \gamma_2 & h(u(k)) \geq \gamma_2, \end{cases} \tag{10.16}$$

where f_b is the frequency of the auditory signals and $h(u(k))$ is a scaling function which converts the range of $u(k)$ to the range of appropriate frequencies of the auditory signal, $\gamma_1 = 400\ Hz$ and $\gamma_2 = 2000\ Hz$.

10.2.3.4 The proposed control mechanism in summary

The structure of the designed control system can be briefly expressed as follows:

1- The control signals are updated synchronously with the HR measurements at a constant rate.

2- An auditory signal (biofeedback) is generated according to the level of the control signal and transmitted whenever the first event occurs.

3- This auditory signal motivates the exercising subject to increase or decrease their pedaling rate and the level of force applied.

4- As a result of varying the pedaling rate during some sampling periods, more than one event can occur. For these events, the same auditory signal as the one transmitted at the first event is sent to the subject. These additional auditory commands help the subject to continue tracking the desired profile.

10.2.3.5 Two novel anti-windup mechanisms

In the case of utilizing an integrator in the controller, if the actuators perform at their limits independently of the system outputs measured, the error will be integrated continuously for a long period before conditions return to normal. Hence, the control effort may increase to a very large value. Thus the PID or ISMC controller may cause large transients while the system actuator saturates. This is referred to as *windup* phenomena [24]. In this work, as the exercising subject plays the role of system actuator and has exercise limitations, windup can similarly occur.

If the exercising subject inadvertently or deliberately chooses not to follow the controller commands, a large tracking error will be generated, and the controller will require a huge effort to compensate for the absent effort. This may result in a large fluctuation in the HR of the subjects if the exercising subject responds to the biofeedback signal. If the HR profile (such as warm up slope θ_w in (10.1) and target HR (HR_e)) are not determined accurately for each subject, considering the maximum work-rate that the subject can do while pedaling continuously, a collapse in the closed-loop mechanism can occur. Even if the target HR and warm up slope are set appropriately according to each subject's condition, the subject may deliberately choose to stop the exercise for a short period.

In conventional control theory, to deal with this problem, a number of anti-windup schemes have been developed, which need to know and use the actuator limitations. Here, however, the subjects' exercise limitations are not easily obtained. Furthermore, it is not always possible to determine that the subject's unresponsiveness is inadvertent or deliberate. In order to address the problem, the HR profile needs to be more dynamic. In other words, where large tracking errors are noted (larger than a specific threshold), the HR profile can be temporarily held constant to prevent large efforts being required of the subjects. We thus revise the HR profile as follows:

$$\theta_w = \begin{cases} \frac{HR_e - HR_n}{T_w} & \text{if } e(k) \leq \delta, \\ 0 & \text{if } e(k) > \delta, \end{cases} \tag{10.17}$$

where $\delta > 0$ denotes a certain threshold. Notice that the warm up duration will then increase to $\bar{T}_w = T_w + T_a$ in which T_a is the summation of all time periods that $e(t) > \delta$ and $\theta_w = 0$. This can avoid any large fluctuation in the HR of the subject. Fig. 10.6 manifests the concept of a dynamic profile during a trial exercise. Here the tracking error threshold is assumed to be $\delta = 10$.

On the other hand, since the designed control system contains an auditory converter (biofeedback generator) it is also possible that both converters (audio frequency and time duration) may reach their limitations, even with the accurate definition of the translating functions t_d and f_b in (10.15) and (10.16), respectively. This can also result in windup phenomena. Note that the velocity form PID control strategy first computes $\Delta u(k)$, see (10.3), which is then integrated via an integrator. This possible windup problem can readily be addressed by assuming $\Delta u(k) = 0$ during the

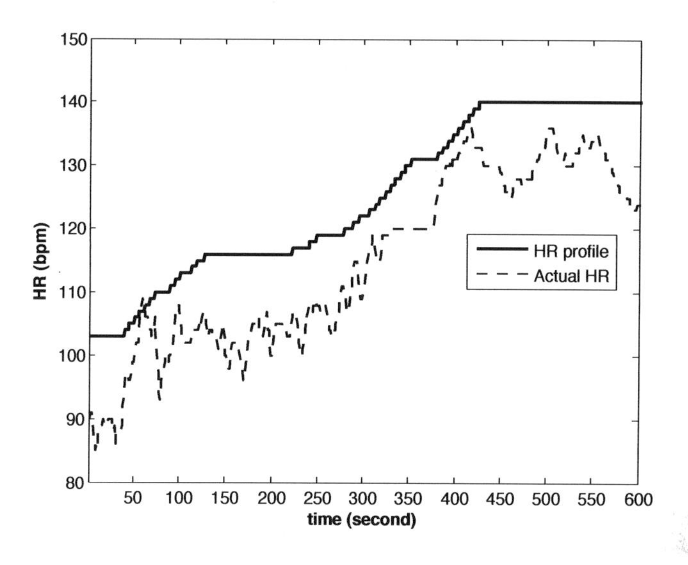

FIGURE 10.6
Dynamic profile mechanism used as anti-windup

saturation. Thus, we define

$$\begin{cases} u(k) = u(k-1) + \Delta u(k), & \text{if } \begin{cases} \lambda_1 < g(u(k)) < \lambda_2 \ \& \\ \gamma_1 < h(u(k)) < \gamma_2 \end{cases} \\ u(k) = u(k-1), & \text{otherwise.} \end{cases} \tag{10.18}$$

Furthermore, we address the possible windup by switching off the discrete-time integrator in the designed ISMC, while the converters perform at their limits independently of the control signals calculated. This can be shown as:

$$\begin{cases} u_{aw}(k) = u_{aw}(k-1), & \text{if converters perform at their limits} \\ u_{aw}(k) = u(k), & \text{otherwise,} \end{cases} \tag{10.19}$$

where u_{aw} denotes the control signal which is given to the auditory converter part.

10.2.3.6 Relay controller strategy

Now we explain the details of implementation of the relay controller designed for tracking the HR profile during cycling. This controller generates a sequence of commands by showing a number of *phrases* on the computer station. Every phrase suggests the desired action required by the subject to achieve the desired HR. The commands are determined based on the error defined in (10.5). The switching of the commands are as follows:

$$u(k+1) = \begin{cases} \text{Apply high intensity effort} & e(k) > \varepsilon_u \\ \text{Apply modest intensity effort} & \varepsilon_l < e(k) \leq \varepsilon_u \\ \text{Maintain current level of effort} & -\varepsilon_l \leq e(k) \leq \varepsilon_l \\ \text{Reduce intensity of effort} & -\varepsilon_u \leq e(k) < -\varepsilon_l \\ \text{Minimize intensity of effort} & e(k) < -\varepsilon_u \end{cases} \quad (10.20)$$

where $\varepsilon_l > 0$ and $\varepsilon_u > 0$ are the given values. Similar to the PID control part, again, the relay controller commands are synchronized to the position of the pedals which are sensed via the reed sensors. In other words, the designed relay controller is also actuator-based event-driven.

10.2.4 Tuning the PID controller gains

This subsection describes the system identification method that we used to make a model describing a system from the measured input-output test data. Participants were requested to complete exercise sessions on separate occasions. In each session, the subject was requested to complete an exercise on the cycle-ergometer with the following setup. In order to excite the system dynamics, we sent a minimum possible biofeedback for 10 seconds to the subjects. During this period the subjects needed to exercise with the minimum possible rate. Then we changed the auditory commands to the maximum case. Using the Matlab identification toolbox, the so-called average transfer function could be obtained. Based on the achieved average transfer function, we optimized PID gains to have a concurrent fast and stable response using the Matlab PID tuner toolbox. This scheme improved the robustness of the PID controller against the inter-individual differences of the HR dynamics. Notice that traditionally PID gains can be tuned using the well-known Zigler-Nichols scheme. However, this tuning method gives different values for different subjects.

10.3 Results

In order to validate the proposed control system, a group of healthy male subjects were asked to exercise on the cycle-ergometer; see [16]. The PID controller parameters were tuned to $K_p = 2.45$, $K_i = 0.02$ and $K_d = 0.06$, which were obtained through the average transfer function of the HR response (see Subsection 10.2.4) and are relatively robust against the inter-individual differences in the dynamic HR response to work rate. A wide range of experiments were carried out on a number of subjects during the cycling exercise to validate the control system. Note that while laboratory conditions were maintained constant for all participants, some uncontrollable factors, such as heat loss, dehydration and humidity, could still influence the results. Hence, the proposed exercise method was repeated twice for each subject. The HR profile was defined in 3 stages; warm up, aerobic training and cool down, where the aerobic

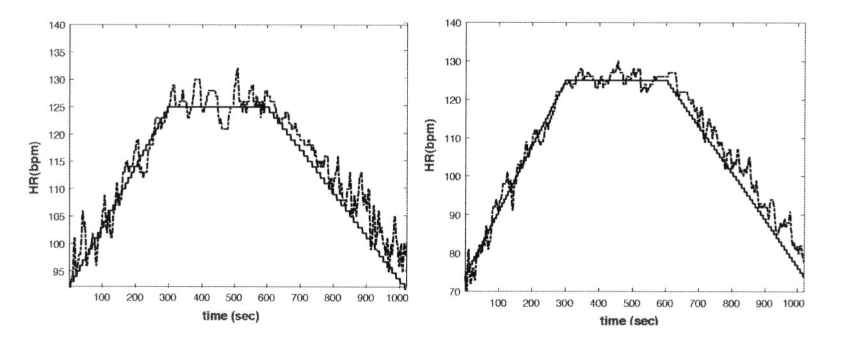

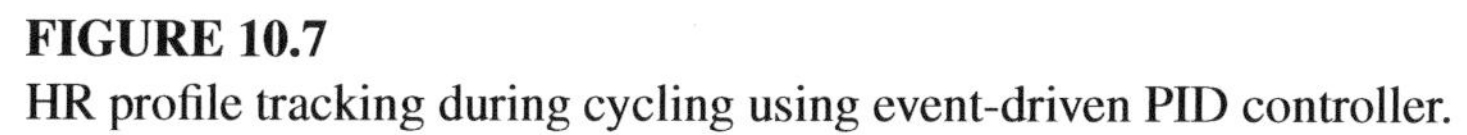

FIGURE 10.7
HR profile tracking during cycling using event-driven PID controller.

training stage lasted 5, 5 and 7 minutes for each subject respectively. Moreover, the reference value of 125 *bpm* was our desired aerobic training HR for all subjects during the exercise stage. The tracking performance of this controller for two subjects is demonstrated in Figs. 10.7. It can be concluded from these results that the designed PID controller can efficiently and robustly drive the subject to track the desired HR profile. For comparison, twelve healthy participants were asked to exercise again, using a conventional PID controller with the same gains given above, and further, using a fixed-rate biofeedback mechanism (subjects were provided with an auditory stimulus every 1 *s* using the last updated control signal; see [25]). Except the controller and bio-feedback rate, the experiments were implemented with the same setup of the previous experiments. The performance of the controller for a subject can be seen in Figs. 10.8. In addition, as another simple control strategy, we have used the relay controller as explained in Section 10.2.3.6, with the following tuned parameters:

$$\varepsilon_l = 3, \ \varepsilon_u = 10.$$

Again, twelve subjects, who were involved in the PID control validation, were requested to perform the cycling exercise using the developed relay controller. This controller generates a sequence of visual commands (phrases on the computer station). The visual commands suggest the desired action required by the subject to achieve the desired HR. The performance of this control strategy for one subject is shown in Fig. 10.9. Lastly, the performance of the ISMC for a subject can be seen in Figs. 10.10. The results illustrate that the novel actuator-based event driven ISMC, designed specifically for cycle ergometer exercises, improves tracking performance and reduces the possibility of the subject's cognitive disengagement from the feedback controller.

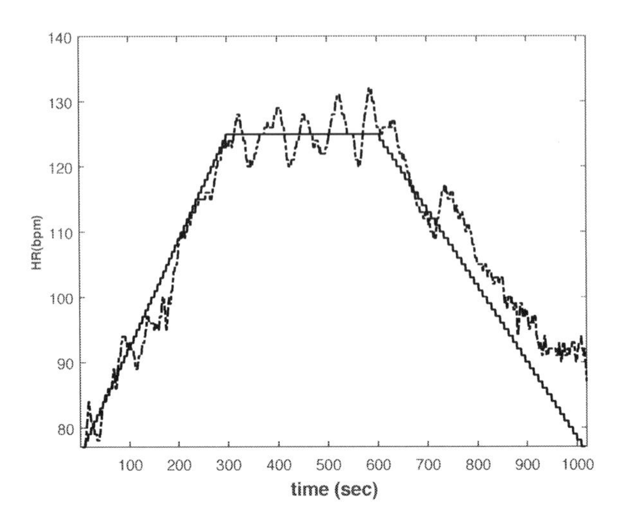

FIGURE 10.8
HR profile tracking during cycling using conventional PID controller and fixed-rate biofeedback

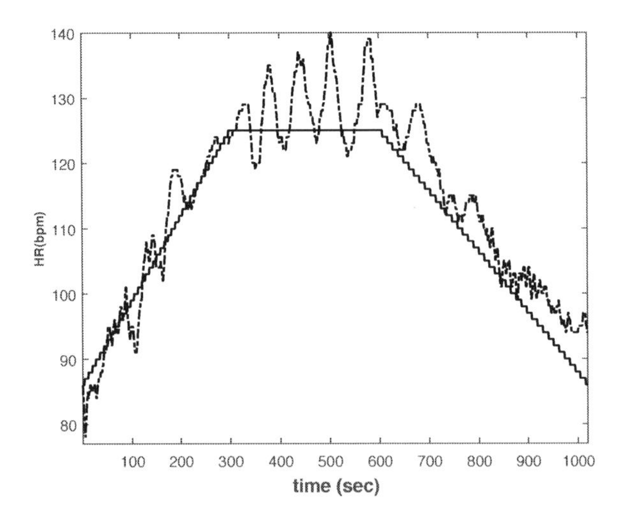

FIGURE 10.9
HR profile tracking during cycling using the relay controller

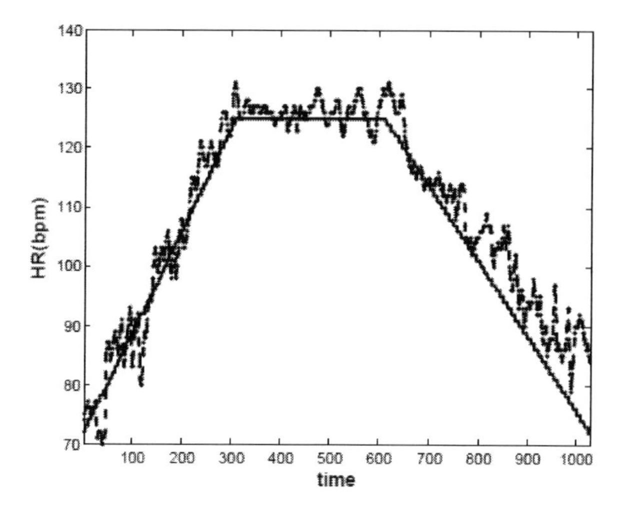

FIGURE 10.10
HR profile tracking during cycling using event-driven ISMC and damped RLS

10.4 Discussion

10.4.1 Necessity of the project

Automated exercise testing systems have become more and more important in several areas such as sport training, medical diagnosis, rehabilitation and analysis of cardiorespiratory kinetics [37, 131]. Moreover, controlling the HR during exercise is of great importance in exercise protocols designed for patients with cardiovascular diseases and those recovering from an infarct or cardiac surgery [34, 69].

Cycle-ergometers are often considered preferable to treadmills, in terms of cost, floor space occupied, level of noise pollution and the risk of falling or injury. Moreover, the upper body motion is usually reduced during cycle-ergometer exercises, compared to the treadmill exercises, and hence, measuring the heart rate (HR) is easier and more reliable [45]. Cycle-ergometers with an externally controllable automatic braking system have been utilized in the literature as an automated exercise system whose mechanism works based on adjusting the resistance of the brake rather than the pedaling rate to change the exercise level [75]. However, the cycle-ergometer with built-in braking system is often expensive and may not be suitable for use at home as a rehabilitation device. Furthermore, our experience is that patients find increasing the workload by increasing the pedaling rate more acceptable than increasing the workload at a constant pedaling rate by increasing the braking effect. In this project, we have alternatively selected an air vane bike that is low cost, and very sturdy. The design is also particularly safe and suitable for use by frail elderly patients. Moreover, the cooling effect of the air vane bike can be regarded as a distinct advantage, see Fig. 10.11. On the other hand, in such a case, the controller signal must be transmitted directly to the subject through some sort of biofeedback. Biofeedback has been used in rehabilitation [52] for more than fifty years. In general, it serves to provide biological information for patients in real-time and to motivate them to implement desired tasks [52]. Biofeedback usually requires the measurement of a subject's biomedical variables and transference to the subject through either direct feedback according to the measured variables, i.e. displaying a numerical value on a wearable device (e.g. watch), or converted feedback, i.e. a mechanism designed to generate an auditory signal, visual display or tactile feedback based on the subject's measured biomedical variables [52, 110, 99].

Existing auditory (or visual) biofeedback mechanisms, e.g. in [25], are not appropriate for this project, as they may lead to a cognitive disengagement of the subject from the feedback controller and possible abrupt changes in the subject's HR reference. We have developed a different mechanism for generating and transmitting biofeedback that needs a novel control system strategy, which can smooth out the abrupt changes in the control signals preventing overshooting in the HR responses, and allowing HR to vary in a more gradual and smoother manner. This smooth variation of HR response is a very important parameter in the automated exercise testing systems developed for the cardiac rehabilitation programs. The proposed control system is discussed in the following subsection.

10.4.2 Discussion of the control system

Most commonly, control theory and engineering consider control systems to be time-driven where continuous-time signals are expressed by their sampled values at a fixed sample frequency. This means that the sampling intervals are assumed to be fixed, hence, the analysis problems can be implemented by utilizing the existing literature on the sampled data systems. The control actions are also assumed to be generated at a frequency equal to the sampling frequency, which is referred to as synchronous in time, or time-driven control.

However, some applications need event-driven controllers [61]. In other words, the controller actions are not synchronous with the discrete-time system sampling frequency [61]. For event-driven controllers, the occurrence of a certain event triggers the controllers actions. As an illustration, consider a motion control system including a motor and an encoder. The encoder pulses happen at regular angular position periods [61]. Upon increasing the speed of the motor, the controller receives pulses from the encoder at a higher rate, which in turn, leads to a higher controller run rate. However, for low velocities, the quantization errors cannot be ignored. One possible method to solve this kind of problem is to use a high-resolution sensor (encoder) which increases the overall system cost. The other alternative method, considered in the literature, is to implement a revision to conventional control algorithms in order to still employ the low-resolution sensors. Two different methods have been suggested in terms of using low-resolution sensor data [61].

1) Utilizing an observer-based framework to estimate the data at synchronous con-

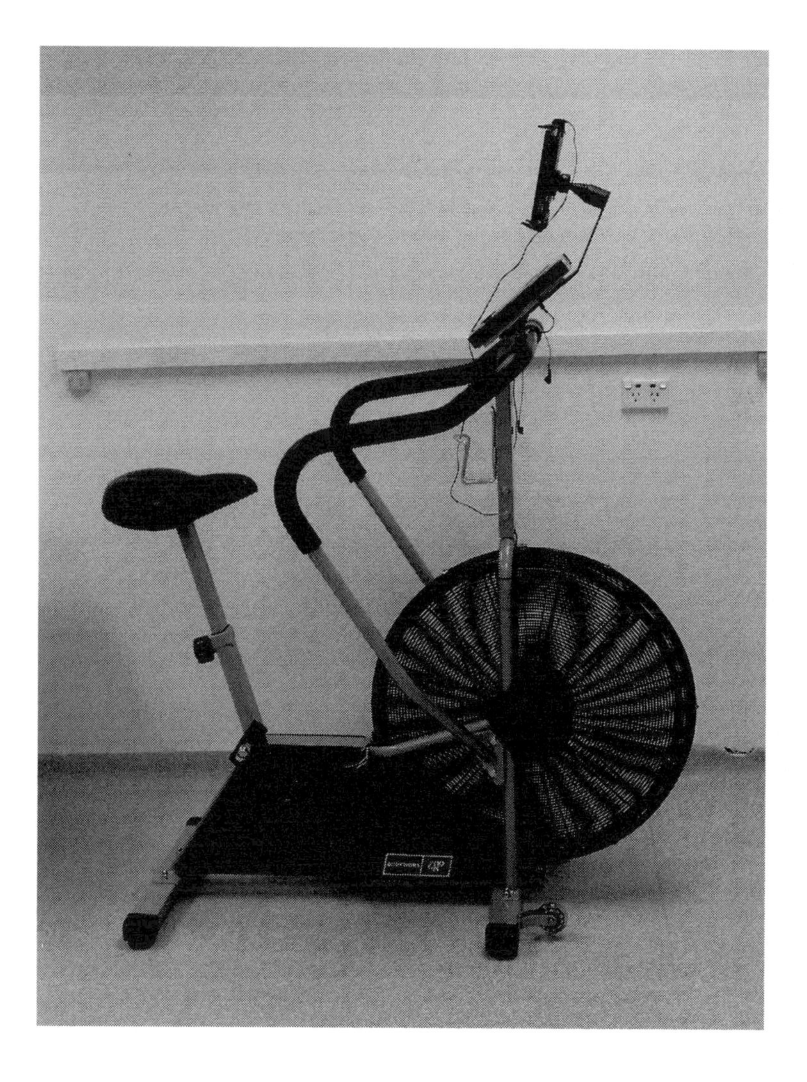

FIGURE 10.11
Cycle-ergometer exercising system

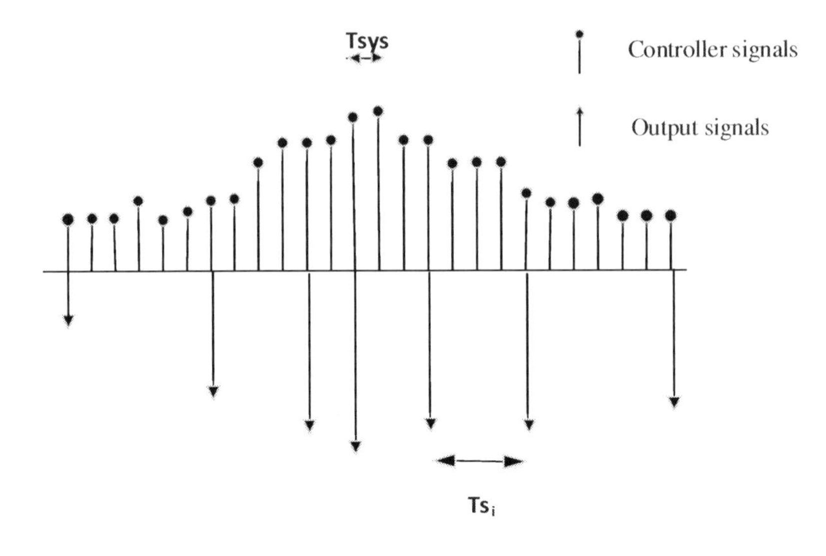

FIGURE 10.12
Mechanism of a sensor-based event driven control system using state observer. T_{s_i} is the varying output sampling rate and T_{sys} is the sampling rate of the system.

troller sample instants, using asynchronous measurement instants. Although these observer-based methods represent better performance compared to conventional control schemes, they have a major drawback, which is the intensive computational burden on the processor, as they need to deal with a time-varying observer-based control system [61], see Fig. 10.12.

2) Defining the models of the plant and the controller in the spatial domain instead of the time domain. Then the controller design can be performed using conventional control theory, see Fig. 10.13.

The sampling rate of the HR transmitted to the program by an ECG or pulse oximeter transducer is typically constant (1 sample per second), however, the rate of sending the commands to the exercising subject varies according to the pedaling rate, see Fig. 10.14. This form of control system, called here an *actuator-based event-driven control system*, is obviously different from the two existing control systems in [61]. Possibly, it can be considered to some degree as a multi-rate system, as the rates of output sampling and control input updating are different, however, since the updating rate of the control input varies according to the pedaling rate of the exerciser, it can also be assumed to be an event-based control system. Nevertheless, our proposed system is not sensor-based event-driven. This is because the aim of this control system is to control the exercise rate through the heart rate and not to control the velocity or the position of a rotary system (DC motor in the aforementioned control systems

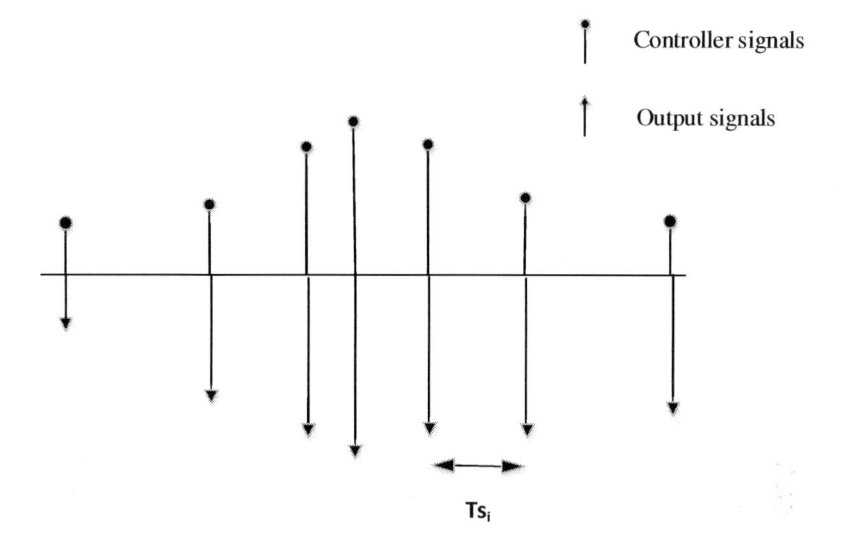

FIGURE 10.13
Mechanism of a sensor-based event driven control system using spatial domain instead of time domain. T_{s_i} is the varying output sampling rate and control signal update rate.

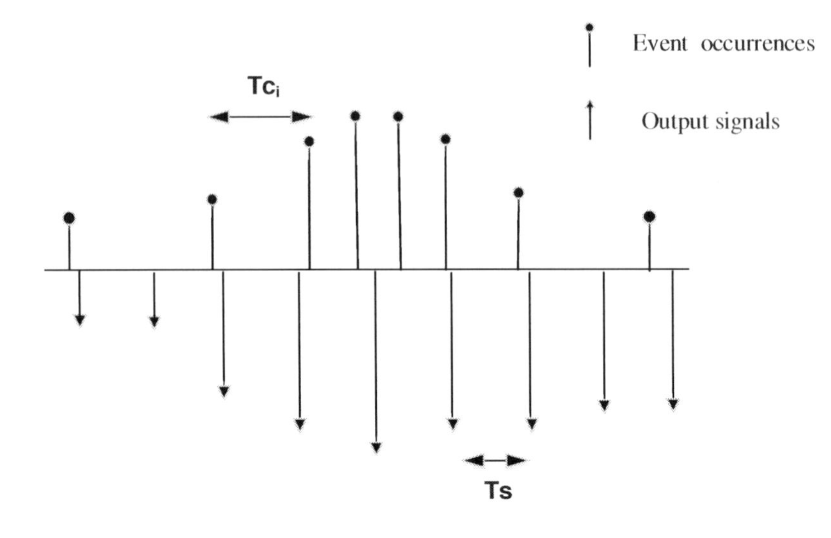

FIGURE 10.14
Mechanism of an actuator-based event driven control system. T_s is the constant output sampling rate and T_{c_i} is the varying event occurrence rate (pedaling rate).

and the pedaling system here) and the HR sensor signal, sampled with a constant rate independent from the pedaling rate, will not trigger the control actions. Fig. 10.1 gives a general view of the asynchronous signals in the designed control system.

The second method explained above for the sensor-based event-driven control systems, which uses the spatial domain of the system model, cannot be adapted to our work. This is because this scheme requires the control signals and sensor (encoder) signals to be synchronous. Furthermore, both methods need the accurate model of the system. Since this work is aimed at developing a non-model-based PID controller, an accurate model of the system is not available.

It should be noted that another kind of event-triggering mechanism with the purpose of reducing the resource utilization of its implementation has been considered [61]. This mechanism updates the control effort only when the error is larger than a specific threshold and holds the control effort while the error is small. To illustrate this, let us consider the following PID controller:

$$u(t) = K_p e(t) + K_i \int_0^t e(\tau)d\tau + K_d \dot{e}(t). \tag{10.21}$$

Using the backward differentiation method we have,

$$\frac{u(k)-u(k-1)}{T_{s_k}} = K_p\frac{e(k)-e(k-1)}{T_{s_k}} + K_i e(k) + K_d\frac{\frac{e(k)-e(k-1)}{T_{s_k}} - \frac{e(k-1)-e(k-2)}{T_{s_{k-1}}}}{T_{s_k}}, \tag{10.22}$$

where T_{s_k} denotes the varying sample time. Then it follows,

$$u(k) = u(k-1) + K_p[e(k)-e(k-1)] + K_i T_{s_k} e(k) + \frac{K_d}{T_{s_k}T_{s_{k-1}}}[T_{s_{k-1}}(e(k)-e(k-1)) - T_{s_k}(e(k-1)-e(k-2))]. \tag{10.23}$$

The controller in (10.23) is suitable as a velocity form event-driven PID controller. Note that we ignore the use of a low-pass filter to deal with the high frequency measurement noise here. The event-triggering mechanism then applies based on the tracking error as

$$\tau_{k+1} = \inf\{t \geq T_{s_m} + \tau_k, \text{ if } |e(t)| \geq \gamma\}, \tag{10.24}$$

where τ_k is the time instant that the k-th controller update is generated. Also, $T_{s_m} > 0$ and $\gamma > 0$ denote the minimum sampling time of the system and the threshold value, respectively. While $e(t) < \gamma$ the control effort is held constant and the sample time T_{s_k} varies. Again, note that this non-uniform updating mechanism requires that no limitation is imposed on the time instant that the control efforts apply. However, in our case, the event triggering mechanism only varies the biofeedback time instants. One alternative is to neglect the triggering criteria in (10.24) and to use the controller in (10.23), which also imposes additional online calculations on the processor. However, it is clear that due to the fixed sampling rate of the HR sensor, in cases where more than one update is required during each sample period, no new HR value, and hence, tracking error is available for the second and later updates. Now three possible methods can be considered, which are as follows:

1. Updating the control signal for each event using the latest available information. However, owing to the existence of the integral action in the control law, this control strategy may be more likely to suffer from *windup* phenomena in the system performance.
2. Designing a control system which utilizes an observer-based framework to estimate first the next actuator event instant (the instant that the left or right pedal passes the reed sensor) and, second, the HR value at asynchronous controller event instants, using synchronous measurement instants. Although this observer-based method may have a better performance compared to the conventional control schemes, it has a major drawback, which is the intensive computational burden on the processor, as it needs to deal with a time-varying observer-based control system.

3. Utilizing a time-driven PID controller in the velocity form such as the one explained in Section 10.2.3.
   An immediate advantage of the last method is that it can assist in preventing the converted control commands (auditory stimulus) from reaching their limitations (see Subsections 10.2.3.3 and 10.2.3.5) and the system working under a saturation situation.

It is also well known that the heart rate response to exercise can be approximated as a first order system, where the time constant is relatively slow [60]. Thus, the variation of the human HR is usually limited in a sampling period (say 1 second). So, even if the controller uses the latest available sensor's information, rather than the actual one, the obtained control signal will be close to the desired one. Moreover, the number of additional events in the sampling period of the HR sensor is not big. With this specific bicycle, we found that the pedaling rate varies in the range $20 \sim 120$ *rpm*, which means that at most, 2 or 3 events may occur in the fixed sampling period. Basically, the additional auditory signals are used here to instruct the subject to retain the proper exercise rate. Although the proposed control scheme here is very simple, our experimental studies (given in this chapter) prove its effectiveness. As opposed to this, the lack of auditory commands for additional events occurring during the HR sampling time, make for an uncomfortable exercising experience for the subject. Fig. 10.1 describes schematically the control system designed for the task.

10.4.3 Discussion of the results

The results in Figs. 10.7-10.9 illustrate that although the relay control method can help the user to track the desired HR profile, the performance of the ISMC/PID PID control and the auditory converter is superior compared to the relay controller. Indeed, compared to the relay controller, the proposed actuator-based event-driven PID controller as well as ISM controller can prevent major overshooting in the HR responses. This may be important, i.e. in cardiac rehabilitation.

Notice that the relay controller is a discontinuous controller which works based on switching functions. Also, the parameters ε_l and ε_u in (10.20) play an important role in the performance of this controller. Smaller values of these parameters may lead to better performance. At the same time, it can also lead to more visual commands (phrases on the computer station) which need to be read and interpreted continuously by the subject. For some subjects, this may contribute to cognitive overload and disengagement.

Set against this, the novel actuator-based event driven control system, designed specifically for cycle ergometer exercises, improves tracking performance and reduces the possibility of the subject's cognitive disengagement from the feedback controller. Furthermore, as can be seen in Fig. 10.7, this control mechanism smoothes out the abrupt variations in the HR profile preventing overshooting in the HR responses, and allows the HR to change in a more gradual manner.

Furthermore, it is found that the fixed-rate biofeedback mechanism (e.g. in [25]) is not a very successful one to instruct the subjects to change their exercise rate according to the controller signal, as the rate of biofeedback is not synchronous with the rate of exercise rate (pedaling rate in this case). Referring to the experiences reported by the exercise participants, this method may also lead to cognitive disengagement of the exercising subject from the controller signals and, consequently, larger overshooting in the HR responses. This can be a major drawback of the control system, especially in cardiac rehabilitation programs. Specifically, during the warm up and cooling down periods (the first and third stages of the HR profile), as the pedaling rate varies repeatedly, it is hard for the exerciser to follow the fixed-rate biofeedback signal and a confusion may occur.

We conclude that the event-driven controller presented in this chapter makes it possible to precisely regulate HR to a predetermined HR profile.

It should also be pointed out that the subjects' HR did not completely follow the HR profile during the cooling down period. This is because the recovery HR of the subjects could not be further reduced since the subjects had already stopped their exercise. In other words, if a subject has a high resting HR which is higher than the reference HR, nothing could be done by the control system to reduce the subject's resting HR. However, in such a case, the cooling down period aims to prevent possible venous pooling and reduced venous return to the heart from an excessively abrupt termination to the exercise.

10.5 Conclusions

In this study, an automated system has been designed to help exercising subjects to track a predetermined HR profile. Three kinds of control methods (ISMC, PID and relay controller), which are simple to design and implement, were deployed and tested in this project. The significance of this study, compared to other published data, is the fact that the biofeedback signals are synchronized with respect to the positions of the pedals. To this end, in the case of the PID or ISMC controller, by adjusting the time duration and the frequency of the auditory signals, the HR of the subjects is forced to track the profile. To implement this idea, a new control system strategy has been designed, termed an *actuator-based event-driven control system*. In addition, for comparison, a simple relay controller and a conventional PID controller using a fixed-rate biofeedback mechanism have been used with their performances compared to the actuator-based event-driven PID control as well as ISMC. Experimental results, which were carried out on male participants, validated the effectiveness of the system.

Bibliography

[1] Khalid Abidi, Jian-Xin Xu, and Yu Xinghuo. On the discrete-time integral sliding-mode control. *IEEE Transactions on Automatic Control*, 52(4):709–715, 2007.

[2] Vincent Acary and Bernard Brogliato. Implicit Euler numerical scheme and chattering-free implementation of sliding mode systems. *Systems & Control Letters*, 59(5):284–293, 2010.

[3] Hassan Adloo, Paknosh Karimaghaee, and Ahad Soltani Sarvestani. An extension of sliding mode control design for the 2-D systems in Roesser Model. In *CDC*, pages 7753–7758, 2009.

[4] S. Amin. Smart grid: Overview, issues and opportunities. advances and challenges in sensing, modeling, simulation, optimization and control. *Euro. Jour. of Cont*, 17(5-6):547–567, 2011.

[5] A. Argha, L. Li, and S. W. Su. A new approach to applying discrete sliding mode control to 2D systems. In *Proc. 52nd IEEE Conference on Decision and Control*, pages 3584–3589, Florence, Italy, Dec. 2013.

[6] A. Argha, L. Li, S. W. Su, and H. Nguyen. Discrete-time sliding mode control for networked systems with random communication delays. In *American Control Conference (ACC), 2015*, pages 6016–6021. IEEE, 2015.

[7] Ahmadreza Argha, Li Li, Steven Su, and Hung Nguyen. Stabilising the networked control systems involving actuation and measurement consecutive packet losses. *IET Control Theory & Applications*, 10(11):1269–1280, 2016.

[8] Ahmadreza Argha, Li Li, and Steven W. Su. H_2-based optimal sparse sliding mode control for networked control systems. *International Journal of Robust and Nonlinear Control*, pages 16–30, 2018. rnc.3852.

[9] Ahmadreza Argha, Li Li, Steven W. Su, and Hung Nguyen. Controllability analysis of the first FM model of 2D systems: A row (column) process. In *2014 IEEE 53rd Annual Conference on Decision and Control (CDC)*, pages 2414–2419. IEEE, 2014.

[10] Ahmadreza Argha, Li Li, Steven W. Su, and Hung Nguyen. Controllability analysis of two-dimensional systems using 1D approaches. *IEEE Transactions on Automatic Control*, 60(11):2977–2982, 2015.

[11] Ahmadreza Argha, Li Li, Steven W. Su, and Hung Nguyen. Robust output-feedback discrete-time sliding mode control utilizing disturbance observer. In *2015 IEEE 54th Annual Conference on Decision and Control (CDC)*, pages 5671–5676. IEEE, 2015.

[12] Ahmadreza Argha, Li Li, Steven W. Su, and Hung Nguyen. H_2-based optimal sparse sliding mode control for networked control systems. In *2016 IEEE 55th Conference on Decision and Control (CDC)*, pages 6826–6831. IEEE, 2016.

[13] Ahmadreza Argha, Li Li, Steven W. Su, and Hung Nguyen. On LMI-based sliding mode control for uncertain discrete-time systems. *Journal of the Franklin Institute*, 353(15):3857–3875, 2016.

[14] Ahmadreza Argha, Li Li, Steven W. Su, and Hung Nguyen. Sparse observer-based sliding mode control for networked control systems. *IFAC-PapersOnLine*, 50(1):12997–13002, 2017.

[15] Ahmadreza Argha, Li Li, and Steven W. Su. Sliding mode stabilisation of networked systems with consecutive data packet dropouts using only accessible information. *International Journal of Systems Science*, 48(6):1291–1300, 2017.

[16] Ahmadreza Argha, Steven W. Su, and Branko G. Celler. Heart rate regulation during cycle-ergometer exercise via event-driven biofeedback. *Medical & Biological Engineering & Computing*, 55(3):483–492, 2017.

[17] Ahmadreza Argha, Steven W. Su, Sangwon Lee, Hung Nguyen, and Branko G. Celler. On heart rate regulation in cycle-ergometer exercise. In *Engineering in Medicine and Biology Society (EMBC), 2014 36th Annual International Conference of the IEEE*, pages 3390–3393. IEEE, 2014.

[18] Ahmadreza Argha, Steven W. Su, Hung Nguyen, and Branko G. Celler. Designing adaptive integral sliding mode control for heart rate regulation during cycle-ergometer exercise using bio-feedback. In *Engineering in Medicine and Biology Society (EMBC), 2015 37th Annual International Conference of the IEEE*, pages 6688–6691. IEEE, 2015.

[19] Ahmadreza Argha, Steven W. Su, Hung Nguyen, and Branko G. Celler. Heart rate regulation during cycle-ergometer exercise via bio-feedback. In *2015 37th Annual International Conference of the IEEE Engineering in Medicine and Biology Society (EMBC)*, pages 4639–4642. IEEE, 2015.

[20] Ahmadreza Argha, Lin Ye, Steven W. Su, Hung Nguyen, and Branko G. Celler. Heart rate regulation during cycle-ergometer exercise using damped parameter estimation method. In *2016 IEEE 38th Annual International Conference of the Engineering in Medicine and Biology Society (EMBC)*, pages 2676–2679. IEEE, 2016.

[21] Ahmadreza Argha, Lin Ye, Steven W. Su, Hung Nguyen, and Branko G. Celler. Real-time modelling of heart rate response during exercise using a novel constrained parameter estimation method. In *2016 IEEE 38th Annual International Conference of the Engineering in Medicine and Biology Society (EMBC)*, pages 2680–2683. IEEE, 2016.

[22] Wilbert S. Aronow. Exercise therapy for older persons with cardiovascular disease. *The American Journal of Geriatric Cardiology*, 10(5):245–252, 2001.

[23] Karl J. Åström and Björn Wittenmark. *Adaptive Control*. Courier Corporation, 2013.

[24] Karl Johan Åström and Tore Hägglund. *Advanced PID control*. ISA-The Instrumentation, Systems, and Automation Society; Research Triangle Park, NC 27709, 2006.

[25] Dur-e-Zehra Baig, Faizan Javed, Andrey V. Savkin, and Branko G. Celler. An adaptive H_∞ control design for exercise-independent human heart rate regulation system. In *2011 9th IEEE International Conference on Control and Automation (ICCA)*, pages 1033–1036. IEEE, 2011.

[26] J. Baillieul and P. Antsaklis. Control and communication challenges in networked real-time systems. *Proceedings of the IEEE*, 95(1):9–28, 2007.

[27] A. Bartoszewicz. Discrete-time quasi-sliding-mode control strategies. *IEEE Transactions on Industrial Electronics*, 45:633–637, 1998.

[28] M. Bisiacco. On the state reconstruction of 2-D systems. *Syst. Control Lett.*, 5(5):347–353, Apr. 1985.

[29] M. Bisiacco. State and output feedback stabilizability of 2-D systems. *IEEE Trans. Circuits Syst.*, CAS-32(12):1246–1254, Dec. 1985.

[30] S. Boyd and L. Vandenberghe. *Convex Optimization*. Cambridge University Press, 2004.

[31] E. J. Candes, M. B. Wakin, and S. P. Boyd. Enhancing sparsity by reweighted ℓ_1 minimization. *Journal of Fourier Analysis and Applications*, 14(5-6):877–905, 2008.

[32] Jeang-Lin Chang. Applying discrete-time proportional integral observers for state and disturbance estimations. *IEEE Transactions on Automatic Control*, 51(5):814–818, 2006.

[33] B. Chen, Y. Niu, and Y. Zou. Sliding mode control for networked systems with markovian jumping parameters. In *Proc.12th International Conference on Control, Automation, Robotics and Vision*, pages 1495–1500, Guangzhou, China, Dec. 2012.

[34] Teddy M. Cheng, Andrey V. Savkin, Branko G. Celler, Steven W. Su, and Lu Wang. Nonlinear modeling and control of human heart rate response during exercise with various work load intensities. *IEEE Transactions on Biomedical Engineering*, 55(11):2499–2508, 2008.

[35] H. H. Choi. Variable structure output feedback control design for a class of uncertain dynamic systems. *Automatica*, 38(2):335–341, 2002.

[36] Gery Colombo, Matthias Joerg, Reinhard Schreier, Volker Dietz, et al. Treadmill training of paraplegic patients using a robotic orthosis. *Journal of Rehabilitation Research and Development*, 37(6):693–700, 2000.

[37] R.A. Cooper, S.M. Horvath, J.F. Bedi, D.M. Drechsler-Parks, and R.E. Williams. Maximal exercise response of paraplegic wheelchair road racers. *Spinal Cord*, 30(8):573–581, 1992.

[38] M. Corless. Stabilization of uncertain discrete-time systems. *Proceedings of the IFAC Workshop on Model Error Concepts and Compensation*, 1985.

[39] Peter J. Diggle. *Time SeriesA Biostatistical Introduction*. Oxford Univ. Press, Oxford, UK, 1990.

[40] C. Edwards. A practical method for the design of sliding mode controllers using linear matrix inequalities. *Automatica*, 40:1761–1769, 2004.

[41] C. Edwards, N. O. Lai, and S. K. Spurgeon. On discrete dynamic output feedback min–max controllers. *Automatica*, 41(10):1783–1790, 2005.

[42] C. Edwards and S. K. Spurgeon. Robust output tracking sliding-mode controller/observer scheme. *International Journal of Control*, 64:967–983, 1996.

[43] C. Edwards and S. K. Spurgeon. *Sliding Mode Control: Theory and Applications*. Taylor and Francis, London, 1998.

[44] M. Fardad, F. Lin, and M. R. Jovanovic. Sparsity-promoting optimal control for a class of distributed systems. In *Proc. the American Control Conference*, pages 2050–2055, San Francisco, CA, USA, 2011.

[45] Gerald F. Fletcher, Gary J. Balady, Ezra A. Amsterdam, Bernard Chaitman, Robert Eckel, Jerome Fleg, Victor F. Froelicher, Arthur S. Leon, Ileana L. Piña, and et al. Rodney, Roxanne. Exercise standards for testing and training a statement for healthcare professionals from the american heart association. *Circulation*, 104(14):1694–1740, 2001.

[46] E. Fornasini and G. Marchesini. Doubly indexed dynamical systems: State-space models and structural properties. *Math. Syst. Theory*, 12(1):59–72, 1976.

[47] E. Fornasini and G. Marchesini. State space realization theory of two-dimensional filters. *IEEE Trans. Autom. Control*, AC-21(4):484–492, Aug. 1976.

[48] E. Fornasini and M.E. Valcher. Controllability and reachability of 2-D positive systems: a graph theoretic approach. *IEEE Trans. Circuits Syst. I, Reg. Papers*, 52(3):576–585, Mar. 2005.

[49] Paolo Frasca, Hideaki Ishii, Chiara Ravazzi, and Roberto Tempo. Distributed randomized algorithms for opinion formation, centrality computation and power systems estimation: A tutorial overview. *European Journal of Control*, 24:2–13, 2015.

[50] K. Galkowski, E. Rogers, and D.H. Owens. Matrix rank based conditions for reachability/controllability of discrete linear repetitive processes. *Linear Algebra and its Applications*, 275:201–224, May 1998.

[51] W. Gao, Y. Wang, and A. Homaifa. Discrete-time variable structure control system. *IEEE Trans. Ind. Electron*, 42:117–122, 1995.

[52] Oonagh M. Giggins, U.M. Persson, and Brian Caulfield. Biofeedback in rehabilitation. *J Neuroeng Rehabil*, 10(1):60, 2013.

[53] D. D. Givone and R. P. Roesser. Minimization of multidimensional linear iterative circuits. *IEEE Trans. Comput.*, C-22(7):673–678, July 1973.

[54] D. D. Givone and R. P. Roesser. Multidimensional linear iterative circuits. *IEEE Trans. Comput.*, C-21(10):1067–1073, Oct. 1972.

[55] C. Godsil and G. Royle. *Algebraic Graph Theory*. Springer, 2001.

[56] G. C. Goodwin, H. Haimovich, D. E. Quevedo, and J. S. Welsh. A moving horizon approach to networked control system design. *IEEE Trans. Autom. Control*, 49(9):1427–1445, 2004.

[57] S. Govindaswamy, S. K. Spurgeon, and T. Floquet. Discrete-time output feedback sliding-mode control design for uncertain systems using linear matrix inequalities. *International Journal of Control*, 84:916–930, 2011.

[58] Ignacio E. Grossmann. Review of nonlinear mixed-integer and disjunctive programming techniques. *Optimization and Engineering*, 3(3):227–252, 2002.

[59] G. Gu. *Discrete-Time Linear Systems: Theory and Design with Applications*. Springer, 2012.

[60] M. Hajek, J. Potuček, and V. Brodan. Mathematical model of heart rate regulation during exercise. *Automatica*, 16(2):191–195, 1980.

[61] W.P.M.H. Heemels, J.H. Sandee, and P.P.J. Van Den Bosch. Analysis of event-driven controllers for linear systems. *International journal of control*, 81(4):571–590, 2008.

[62] G. Herrmann, S. K. Spurgeon, and C. Edwards. A robust sliding-mode output tracking control for a class of relative degree zero and nonminimum phase plants: A chemical process. *International Journal of Control*, 72:1194–1209, 2001.

[63] J. Hespanha, P. Naghshtabrizi, and Y. Xu. A survey of recent results in networked control systems. *Proceedings of the IEEE*, 95(1):138–162, 2007.

[64] D. W. C. Ho and G. Lu. Robust stabilization for a class of discrete-time nonlinear systems via output feedback: The unified lmi approach. *Int. J. Control*, 76:105–115, 2003.

[65] Kou-Cheng Hsu. Decentralized variable-structure control design for uncertain large-scale systems with series nonlinearities. *International Journal of Control*, 68(6):1231–1240, 1997.

[66] Olivier Huber, Vincent Acary, Bernard Brogliato, and Franck Plestan. Discrete-time twisting controller without numerical chattering: analysis and experimental results with an implicit method. In *53rd Annual Conference on Decision and Control (CDC)*, pages 4373–4378. IEEE, 2014.

[67] S. Hui and S. H. Zak. On discrete-time variable structure sliding mode control. *Systems and Control Letters*, 38:283–288, 1999.

[68] John Y. Hung, Weibing Gao, and James C. Hung. Variable structure control: a survey. *IEEE Transactions on Industrial Electronics*, 40(1):2–22, 1993.

[69] Kenneth J. Hunt, Simon E. Fankhauser, and Jittima Saengsuwan. Identification of heart rate dynamics during moderate-to-vigorous treadmill exercise. *Biomedical Engineering Online*, 14(1):117, 2015.

[70] Uri Itkis. *Control Systems of Variable Structure*. Halsted Press, 1976.

[71] T. Iwasaki and F. Skelton. Linear quadratic suboptimal control with static output feedback. *Systems and Control Letters*, 23:421–430, 1994.

[72] T. Jiaa, Y. Niu, and Y. Zoua. Sliding mode control for stochastic systems subject to packet losses. *Information Sciences*, 217:117–126, 2012.

[73] T. Kaczorek. *Two-Dimensional Linear Systems*. Springer-Verlag, 1985.

[74] T. Kaczorek. Local controllability, reachability, and reconstructibility of the general singular model of 2-D systems. *IEEE Trans. Autom. Control*, 37(10):1527–1530, Oct. 1992.

[75] Toru Kawada, Yasuhiro Ikeda, Hiroshi Takaki, Masaru Sugimachi, Osamu Kawaguchi, Toshiaki Shishido, Takayuki Sato, Wataru Matsuura, Hiroshi Miyano, and Kenji Sunagawa. Development of a servo-controller of heart rate using a cycle ergometer. *Heart and vessels*, 14(4):177–184, 1999.

[76] Toru Kawada, Genshiro Sunagawa, Hiroshi Takaki, Toshiaki Shishido, Hiroshi Miyano, Hiroshi Miyashita, Takayuki Sato, Masaru Sugimachi, and Kenji Sunagawa. Development of a servo-controller of heart rate using a treadmill. *Japanese Circulation Journal*, 63(12):945–950, 1999.

[77] H. K. Khalil. *Nonlinear Systems, 3rd Edition*. Prentice Hall, New York, 2002.

[78] J. Klamka. *Controllability of Dynamical Systems*. Kluwer, Dordrecht, 1991.

[79] J. Klamka. Controllability of 2D systems. In *Proc. Fourth International Workshop on Multidimensional Systems*, pages 199–206, Wuppertal, Germany, July 2005.

[80] I. V. Kolmanovsky and T. L. Maizenberg. Optimal control of continuous-time linear systems with a time-varying random delay. *Syst. Control Lett.*, 44:119–126, 2001.

[81] P. Korondi, H. Hashimoto, and V. Utkin. Direct torsion control of flexible shaft in an observer-based discrete-time sliding mode. *IEEE Trans. Ind. Electron.*, 45:291–296, 1998.

[82] A.J. Koshkouei and A. S. I. Zinober. Sliding mode control of discrete-time systems. *Journal of Dynamic Systems, Measurement, and Control*, 122:793–802, 2000.

[83] N. O. Lai, C. Edwards, and S. K. Spurgeon. Discrete output feedback sliding-mode control with integral action. *Int. J. Robust Nonlinear Control*, 16:21–43, 2006.

[84] E.P. Lambert. *Process Control Applications of Long-Range Prediction*. PhD thesis, University of Oxford, 1987.

[85] S. M. Lee and B. H. Lee. A discrete-time sliding mode controller and observer with computation delay. *Control Engineering Practice*, 7:2943–2955, 1999.

[86] L. Li, V.A. Ugrinovskii, and R. Orsi. Decentralized robust control of uncertain Markov jump parameter systems via output feedback. *Automatica*, 43:1932–1944, 2007.

[87] S. Li, J. Yang, W.-h. Chen, and X. Chen. *Disturbance observer-based control: methods and applications*. CRC Press, 2014.

[88] F. Lin, M. Fardad, and M. Jovanovic. Augmented Lagrangian approach to design of structured optimal state feedback gains. *IEEE Trans. Autom. Control*, 56(12):2923–2929, 2011.

[89] Merid Lješnjanin, Branislava Draženović, Čedomir Milosavljević, and Boban Veselić. Disturbance compensation in digital sliding mode. In *International Conference on Computer as a Tool (EUROCON)*, pages 1–4. IEEE, 2011.

[90] J. Löfberg. YALMIP: A toolbox for modeling and optimization in MATLAB. In *CCA/ISIC/CACSD*, September 2004.

[91] X. Luan, P. Shi, and F. Liu. Stabilization of networked control systems with random delays. *IEEE Trans. Ind. Elec.*, 58(9):4323–4330, 2011.

[92] M. S. Mahmoud and A. Qureshi. Decentralized sliding-mode output-feedback control of interconnected discrete-delay systems. *Automatica*, 48(5):808–814, 2012.

[93] J. Manela. *Deterministic control of uncertain linear discrete and sampled-data systems.* Ph.D. Thesis, University of California, Berkeley, 1985.

[94] D.J. Mersy. Health benefits of aerobic exercise. *Postgraduate Medicine*, 90(1):103–7, 1991.

[95] C. Milosavljević. General conditions for the existence of a quasi-sliding mode on the switching hyperplane in discrete variable structure systems. *Automation and Remote Control*, 3:36–44, 1985.

[96] Cedomir Milosavljevic, Branislava Perunicic-Drazenovic, and Boban Veselic. Discrete-time velocity servo system design using sliding mode control approach with disturbance compensation. *IEEE Transactions on Industrial Informatics*, 9(2):920–927, 2013.

[97] D. Mitić and C. Milosavljević. Sliding mode based generalised minimum variance control with $o(t^3)$ accuracy. *In Proceedings of the 7th International Workshop on VSS, University of Sarajevo, Bosnia and Herzagovina*, page 6976, 2002.

[98] G. Monsees. *Discrete-time sliding mode control.* Ph.D. Thesis, Delft University of Technolog, The Netherlands, 2002.

[99] R. Montoya, P.H. Dupui, B. Pages, and P. Bessou. Step-length biofeedback device for walk rehabilitation. *Medical & Biological Engineering & Computing*, 32(4):416–420, 1994.

[100] J. Nilsson, B. Bernhardsson, and B. Wittenmark. Stochastic analysis and control of real-time systems with random time delays. *Automatica*, 34(1):57–64, 1998.

[101] Y. Niu and D. W. C. Ho. Design of sliding mode control subject to packet losses. *IEEE Transactions on Automatic Control*, 55:2623–2628, 2010.

[102] Y. Niu, D. W. C Ho, and J. Lam. Robust integral sliding mode control for uncertain stochastic systems with time-varying delay. *Automatica*, 41:873–880, 2005.

[103] Y. Niu, J. Lam, X. Wang, and D. W. C. Ho. Observer-based sliding mode control for nonlinear state-delayed systems. *International Journal of Systems Science*, 35(2):139–150, 2004.

[104] K. Ogata. *Modern Control Engineering*. Prentice-Hall Inc., 1997.

[105] Y. Pan and K. Furuta. VSS controller design for discrete-time systems. *Control-Theory Adv. Technol*, 10(4):669–687, 1994.

[106] Michele Paradiso, Stefano Pietrosanti, Stefano Scalzi, Patrizio Tomei, and Cristiano Maria Verrelli. Experimental heart rate regulation in cycle-ergometer exercises. *IEEE Transactions on Biomedical Engineering*, 60(1):135–139, 2013.

[107] Ronald J. Patton, Chandrasekhar Kambhampati, Alessandro Casavola, Ping Zhang, Steven Ding, and Dominique Sauter. A generic strategy for fault-tolerance in control systems distributed over a network. *European Journal of Control*, 13(2):280–296, 2007.

[108] K. B. Petersen, M. S. Pedersen, and et al. The matrix cookbook. *Technical University of Denmark*, 7:15, 2008.

[109] I. R. Peterson. A stabilization algorithm for a class of uncertain linear systems. *Systems Control Lett.*, 8:351–357, 1987.

[110] J.S. Petrofsky. New algorithm to control a cycle ergometer using electrical stimulation. *Medical & Biological Engineering & Computing*, 41(1):18–27, 2003.

[111] W.A. Porter and J.L. Aravena. 1-D model for m-D processes. *IEEE Trans. Circuits Syst.*, CAS-31(8):742–745, Aug. 1984.

[112] A. Qureshi and M. A. Abido. Decentralized discrete-time quasi-sliding mode control of uncertain linear interconnected systems. *International Journal of Control, Automation and Systems*, 12(2):349–357, 2014.

[113] A. Ray. Output feedback control under randomly varying distributed delays. *J. Guid. Control Dyna.*, 17(4):701–711, 1994.

[114] Mohammad Razeghi-Jahromi and Alireza Seyedi. Stabilization of networked control systems with sparse observer-controller networks. *IEEE Transactions on Automatic Control*, 60(6):1686–1691, 2015.

[115] Robert A. Robergs and Roberto Landwehr. The surprising history of the "hrmax= 220-age" equation. *J Exerc Physiol*, 5(2):1–10, 2002.

[116] Asif Şabanoviç. Variable structure systems with sliding modes in motion control-a survey. *IEEE Transactions on Industrial Informatics*, 7(2):212–223, 2011.

[117] S. Z. Sapturk, Y. Istefanopulous, and O. Kaynak. On the stability of discrete-time sliding mode control systems. *IEEE Transactions on Automatic Control*, 32:930–932, 1987.

[118] L. Schenato. To zero or to hold control inputs with lossy links? *IEEE Trans. Autom. Control*, 54(5):1093–1099, May 2009.

[119] L. Schenato, B. Sinopoli, M. Franceschetti, K. Poolla, and S. S. Sastry. Foundations of control and estimation over lossy networks. *Proc. IEEE*, 95(1):163–187, Jan. 2007.

[120] Simone Schuler, Ping Li, James Lam, and Frank Allgöwer. Design of structured dynamic output-feedback controllers for interconnected systems. *International Journal of Control*, 84(12):2081–2091, 2011.

[121] Simone Schuler, Ulrich Münz, and Frank Allgöwer. Decentralized state feedback control for interconnected process systems. In *IFAC Symposium on Advanced Control of Chemical Processes, Furama Riverfront, Singapore*, pages 1–10, 2012.

[122] Simone Schuler, Ulrich Münz, and Frank Allgöwer. Decentralized state feedback control for interconnected systems with application to power systems. *Journal of Process Control*, 24(2):379–388, 2014.

[123] M. Sebek, M. Bisiacco, and E. Fornasini. Controllability and reconstructibility conditions for 2-D systems. *IEEE Trans. Autom. Control*, 33(5):496–499, May 1988.

[124] M. Shafiee and P. Wellstead. Stability analysis of 2-D systems using the wave advance model with normal matrices. In *Proc. UKACC Int. Conf. Control*, pages 1611–1616, London, UK, Sep. 1998.

[125] D. Siljak. *Decentralized Control of Complex Systems*. Dover Publications, 2012.

[126] B. Sinopoli, L. Schenato, M. Franceschetti, K. Poolla, M. I. Jordan, and S. S. Sastry. Kalman filtering with intermittent observations. *IEEE Trans. Autom. Control*, 49(9):1453–1463, Sep. 2009.

[127] S. K. Spurgeon. Hyperplane design techniques for discrete-time variable structure control systems. *International Journal of Control*, 55(2):445–456, 1992.

[128] D. Srinivasagupta, H. Schattler, and B. Joseph. Time-stamped model predictive control: An algorithm for control of processes with random delays. *Comput. Chem. Eng.*, 28(8):1337–1346, 2004.

[129] Steven W. Su, Shoudong Huang, Lu Wang, Branko G. Celler, Andrey V. Savkin, Ying Guo, and Teddy Cheng. Nonparametric Hammerstein model based model predictive control for heart rate regulation. In *29th Annual International Conference of the IEEE Engineering in Medicine and Biology Society, EMBS 2007.*, pages 2984–2987. IEEE, 2007.

[130] Steven W. Su, Shoudong Huang, Lu Wang, Branko G. Celler, Andrey V. Savkin, Ying Guo, and Teddy M. Cheng. Optimizing heart rate regulation for safe exercise. *Annals of Biomedical Engineering*, 38(3):758–768, 2010.

[131] Steven W. Su, Lu Wang, Branko G. Celler, and Andrey V. Savkin. Heart rate control during treadmill exercise. In *27th International Conference of the IEEE Engineering in Medicine and Biology Society, IEEE-EMBS 2005*, pages 2471–2474. IEEE, 2006.

[132] Steven W. Su, Lu Wang, Branko G. Celler, Andrey V. Savkin, and Ying Guo. Identification and control for heart rate regulation during treadmill exercise. *IEEE Transactions on Biomedical Engineering*, 54(7):1238–1246, 2007.

[133] W. Su, S.V. Drakunov, and Ü. Özgüner. An $O(T^2)$ boundary layer in sliding mode for sampled-data systems. *IEEE Transactions on Automatic Control*, 45:482–485, 2000.

[134] C. Y. Tang and E. Misawa. Discrete variable structure control for linear multivariable systems. *Journal of Dynamic Systems, Measurement, and Control*, 122:783–792, 2000.

[135] C. Y. Tang and E. A. Misawab. Sliding surface design for discrete vss using lqr technique with a preset real eigenvalue. *Systems and Control Letters*, 45(1):1–7, Jan, 2002.

[136] K. C. Toh, M.J. Todd, R.H. Ttnc, and R. H. Tutuncu. SDPT3 - a MATLAB software package for semidefinite programming. *Optimization Methods and Software*, 11:545–581, 1998.

[137] N. Tsai and A. Ray. Stochastic optimal control under randomly varying delays. *Int. J. Control*, 69(5):1179–1202, 1997.

[138] V. I. Utkin. *Sliding Modes in Control Optimization, Communications and Control Engineering Series*. Springer-Verlag, London, 1992.

[139] F. Wang and D. Liu. *Networked Control Systems: Theory and Applications*. Springer, 2008.

[140] X. Wang and M. Lemmon. Event-triggering in distributed networked control systems. *IEEE Transactions on Automatic Control*, 56(3):586–601, 2011.

[141] Z. Wang and F. Yang. Robust filtering for uncertain linear systems with delayed states and outputs. *IEEE Trans. Circuits Syst. I*, 49(1):125–130, 2002.

[142] Junli Wu, Hamid Reza Karimi, and Peng Shi. Network-based H_∞ output feedback control for uncertain stochastic systems. *Information Sciences*, 232:397–410, 2013.

[143] L. Wu and H. Gao. Sliding mode control of two-dimensional systems in Roesser model. *Control Theory & Applications, IET*, 2(4):352–364, 2008.

[144] X.-G. Yan, C. Edwards, and S. K. Spurgeon. Decentralized robust sliding mode control for a class of nonlinear interconnected systems by static output feedback. *Automatica*, 40(4):613–620, Apr. 2004.

[145] X.-G. Yan, S. K. Spurgeon, and C. Edwards. Decentralised sliding mode control for nonminimum phase interconnected systems based on a reduced-order compensator. *Automatica*, 42(10):1821–1828, Oct. 2006.

[146] F. W. Yang, Z. D. Wang, Y. S. Hung, and M. Gani. H_∞ control for networked systems with random communication delays. *IEEE Trans. Autom. Control*, 51(3):511–518, 2006.

[147] H. Yang, Y. Xia, and P. Shi. Observer-based sliding mode control for a class of discrete systems via delta operator approach. *Journal of the Franklin Institute*, 347(7):1199–1213, 2010.

[148] Kohzoh Yoshino, Kimihiro Adachi, Keiko Ihochi, and Katsunori Matsuoka. Modeling effects of age and sex on cardiovascular variability responses to aerobic ergometer exercise. *Medical & Biological Engineering & Computing*, 45(11):1085–1093, 2007.

[149] K. David Young and Umit Ozguner. Sliding mode: Control engineering in practice. In *Proceedings of the 1999 American Control Conference*, volume 1, pages 150–162. IEEE, 1999.

[150] A. Zecevic and D. Siljak. Global low-rank enhancement of decentralized control for large-scale systems. *IEEE Transactions on Automatic Control*, 50(5):740–744, 2005.

[151] Bao-Lin Zhang and Qing-Long Han. Network-based modelling and active control for offshore steel jacket platform with tmd mechanisms. *Journal of Sound and Vibration*, 333(25):6796–6814, 2014.

[152] Bao-Lin Zhang, Qing-Long Han, Xian-Ming Zhang, and Xinghuo Yu. Sliding mode control with mixed current and delayed states for offshore steel jacket platforms. *IEEE Transactions on Control Systems Technology*, 22(5):1769–1783, 2014.

[153] Bao-Lin Zhang, Li Ma, and Qing-Long Han. Sliding mode H_∞ control for offshore steel jacket platforms subject to nonlinear self-excited wave force and external disturbance. *Nonlinear Analysis: Real World Applications*, 14(1):163–178, 2013.

[154] Jinhui Zhang, Yuanqing Xia, and Peng Shi. Design and stability analysis of networked predictive control systems. *IEEE Transactions on Control Systems Technology*, 21(4):1495–1501, 2013.

[155] W. Zhang, M. Branicky, and S. Phillips. Stability of networked control systems. *IEEE Control Systems*, 21(1):84–99, 2001.

Index

ℓ_1 minimization algorithm, 140
ℓ_1 norm, 127, 148
$\mathscr{H}_2$ control, 139–142, 145–149
$\mathscr{H}_2$ cost, 146, 152, 153
$\mathscr{H}_2$ performance, 147, 148
$\mathscr{H}_2$ state feedback, 146

Actuator-based event-driven control system, 187, 188, 191
Air vane bike, 192, 203
Anti-windup, 189, 197
Asymptotic convergence, 4
Asymptotically, 2, 3, 147
Auditory converter, 196–199, 210
Auditory signal, 189
Auditory stimulus, 189, 193, 196, 197, 200, 205, 209–211

Bernoulli sequence, 71, 81, 87, 89–91, 103
Biofeedback, 193, 194, 197–200, 203, 205, 209–211
Boundary layer, 27, 31, 34, 41, 42, 48, 51, 52, 54, 55, 61, 62, 64, 70, 74, 81, 93, 103, 115, 120–122, 133
Boundary region, 10
Boundedness, 31, 34, 37, 40, 51, 52, 55, 59, 64, 66, 70, 73, 74, 77, 87, 92, 102, 123, 124

Chattering, 4, 23
Controllability, 166, 169, 170, 172–181, 183
Coordinate transformation, 21, 140, 147, 163
Cycle-ergometer, 187–189, 191–193, 199, 200, 203

Decentralized SMC, 114, 156
Decentralized topology, 113
Distributed topology, 113
Disturbance estimate, 55, 119, 196
Disturbance observer, 28, 32, 40, 49, 52, 53, 65, 70, 71, 115

Equivalent control, 4, 54, 114, 120, 146, 147, 195
Equivalent controller, 54
Euclidean norm, 2
Event-driven control system, 189
Exogenous disturbance, 2, 28, 31–34, 36, 39, 48, 49, 51–53, 55, 59, 66, 70, 73, 74, 80, 81, 86, 92, 93, 99, 101, 102, 115, 116, 118, 121, 122, 135, 145
External disturbance, 20

Fornasini and Marchesini model, 159
Fully distributed, 113, 114, 121, 131–135

Haskell and Fox formula, 193
Heart rate regulation, 187–189
HR profile, 187, 211
HR response dynamics identification, 189

Ideal sliding, 21, 27, 30, 54, 73, 91, 119, 142, 164
Infinite switching frequency, 1
Integral sliding mode control, 187–189
Interconnected system, 113–115, 128, 144, 147, 150

Large scale system, 115, 126
LQR, 2, 22, 114, 151, 164–166
Luenberger observer, 52, 115

Lyapunov function, 4, 62, 73, 159

Matched uncertainty, 2, 27, 28, 49, 114, 141
Mismatched uncertainty, 34, 74
Model identification, 194
Multi-input system, 15
Multidimensional systems, 159

Networked control systems, 69, 87, 113

Observability, 169, 170
Observable, 52, 55, 89, 118, 121
Original coordinate, 21, 164
Original coordinates, 22
Output feedback, 51, 54, 62, 66, 87, 113–115, 140, 146, 149–151, 153

Phase portrait, 19
PID, 188, 189, 191, 193, 197–200, 202, 208–211
Pole placement, 22, 164
Proportional Integral Observer, 52

Reachability condition, 4, 27, 69, 143
Reconstructibility, 169
Recursive Least Squares, 189, 195
Reduced order system, 22, 147
Regular form, 2, 20, 141, 159, 163, 164
Reweighted ℓ_1 norm, 140, 152, 153, 155
Robust control, 2, 31, 70, 74, 120
Robustness, 130, 136, 140, 199
Roesser Model, 159
Root Mean Square, 134

Sample-and-hold, 1
Sampling, 1
Sampling frequency, 205
Sensor-based event-driven control system, 188
Sign function, 4
Signum function, 38
Single-input system, 15, 140
Singular values, 53
Sliding matrix, 22, 36, 48, 59, 165
Sliding surface, 15, 20–22, 27–29, 32, 34, 35, 39, 42, 51, 52, 54–56, 61, 62, 64, 71, 73, 74, 81, 86, 88, 93, 103, 114, 115, 120, 122, 133, 137, 139–141, 143, 145, 147, 156, 164, 169, 195
Sparsely distributed, 113–115, 121, 126, 139, 144
Sparsification, 128, 139, 140
Stability analysis, 28, 56, 69, 93, 122, 147
State feedback, 88, 114, 151–153, 164
Switching function, 6, 7, 10, 12, 13, 17, 28, 49, 52, 54, 64, 69–71, 86, 102, 142, 144, 147, 164, 210
System identification, 199

Tracking error, 195–198, 209
Trade off, 4, 134
Transfer function, 145, 148, 199, 200

Ultimate boundedness, 28, 86, 103
Unmatched uncertainty, 2

Wave Advanced Model, 159
Weighted ℓ_1 norm, 127, 140, 148